Presented To
Cedar Mill Community Library

In memory of
Dr. James B. Horton, Jr.

FIRE
SEASON

FIRE
SEASON

Field Notes from a Wilderness Lookout

PHILIP CONNORS

An Imprint of HarperCollinsPublishers

Author's note: Some names of places and people have been changed in an effort to protect their innocence.

Grateful acknowledgment is made to the following for permission to reprint copyrighted material:

Oxford University Press for excerpts from *A Sand County Almanac and Sketches Here and There* by Aldo Leopold, copyright © 1968 by Oxford University Press. Reprinted by permission of Oxford University Press.

The University of Chicago Press for excerpts from *Young Men and Fire* by Norman Maclean, copyright © 1992 by the University of Chicago. Reprinted by permission of the University of Chicago Press.

SLL/Sterling Lord Literistic for excerpts from an unpublished fire lookout diary by Jack Kerouac, copyright © 1956 by Jack Kerouac. Reprinted by permission of SLL/Sterling Lord Literistic, Inc.

HarperCollins books may be purchased for educational, business, or sales promotional use. For information please write: Special Markets Department, HarperCollins Publishers, 10 East 53rd Street, New York, NY 10022.

Small portions of this book appeared in different form in *n+1*, *The Nation*, *The Paris Review*, *The Indiana Review*, and the anthology *State by State*, the editors of which are gratefully acknowledged.

FIRST EDITION

Designed by Mary Austin Speaker

Library of Congress Cataloging-in-Publication Data has been applied for.

ISBN 978-0-06-185936-6

11 12 13 14 15 OV/RRD 10 9 8 7 6 5 4 3 2 1

For Martha

CONTENTS

FIRE
SEASON

PROLOGUE

UNTIL ABOUT FIFTEEN YEARS AGO I thought fire lookouts had gone the way of itinerant cowboys, small-time gold prospectors, and other icons of an older, wilder West. Then a friend of mine named Mandijane asked for my mailing address in Missoula, Montana, where we were both students in print journalism—one of the least timely courses of study in the history of higher education, though we couldn't have known that at the time. M.J. said she'd soon have a lot of time to write letters. When spring exams were over, she'd be off to New Mexico to watch for fires.

I was intrigued, and more than a little envious; M.J.'s letters did not disappoint. She was posted in the middle of the Gila National Forest, on the edge of the world's first designated wilderness, 130 miles north of the border with Mexico. On Loco Mountain, she said, not a single man-made light could be seen after dark. She lived in her lookout tower, a twelve-by-twelve-foot room on stilts. The nearest grocery store was five miles by pack trail and eighty-five more by mountain road. Over the course of four months she had fewer than twenty visitors—hunters on horseback, mainly, and a few adventurous hikers. The romance in those letters was almost unimaginable.

1

For years our paths diverged, though we always kept in touch by letter. I left school for New York and lucked into a job with the *Wall Street Journal.* Her continuing adventures took her to Ghana, Costa Rica, and Argentina. One spring she wrote to say she was back in the States for another summer gig in the Gila, this time at a different tower forty miles southeast of Loco Mountain. She knew I was busy, tied to a desk in New York, but suggested I take a vacation and come see the country, for a few days at least.

I needed no further urging. I'd already hustled too long and for no good purpose in the city, and when I finally looked out on that country with two dozen mountain ranges I couldn't name, more mountains than a person could hope to explore on foot in a lifetime, I guess you could say I fell in love at first sight. And what a sight it was: a stretch of country larger than the state of Maryland, nearly 20,000 square miles of desert and forest, sky island mountain chains in three states and two countries. In the afternoons, when M.J. sat in the tower keeping watch, I hiked through old-growth fir and massive groves of quaking aspen. I was unaware of it at the time, but those aspen had grown back in the scar of what was, for almost half a century, the biggest fire on record in the Southwest: the McKnight Fire of 1951, which burned 50,000 acres along the slopes of the Black Range. Much of the fire crowned in mature timber, creating a massive stand replacement—the death of one or several tree species and their total succession by others. Though I could not see it yet, I'd been seduced on my walks by that fire, or at least by the effects of the fire, the beauty of the forest created in its wake.

Around our own little bonfire under starlight, M.J. told me she'd grown antsy in the lookout. She wanted to get out on a fire, inhale the smoke, feel the heat of the flames—and make some bigger money, overtime and hazard pay. Her boss was game, she said, if she could find him someone reliable to take over fire watch. By the time I had to hike out and head home, I'd talked myself into her job. She'd vouch for my backwoods bona fides—atrophied after four years in the city—and I'd fly to New York, offer two weeks' notice at work, and be back before the moon was full again. I knew almost nothing about being a lookout except what I'd read in books, but what I'd read seemed promising. "It doesn't take much in the way of body and mind to be a lookout," Norman Maclean had written. "It's mostly soul."

Since that first summer I've returned each succeeding year to sit 10,000 feet above sea level and watch for smoke. Most days I can see a hundred miles in all directions. On clear days I can make out mountains 180 miles away. To the east stretches the valley of the Rio Grande, cradled by the desert: austere, forbidding, dotted with creosote shrubs, and home to a collection of horned and thorned species evolved to live in a land of scarce water. To the north and south, along the Black Range, a line of peaks rises and falls in timbered waves; to the west, the Rio Mimbres meanders out of the mountains, its lower valley verdant with grasses. Beyond it rise more mesas and mountains: the Diablos, the Jerkies, the Mogollons. A peaceable kingdom, a wilderness in good working order—and my job to sound the alarm if it burns.

Having spent eight summers in my little glass-walled perch, I have an intimate acquaintance with the look and feel of the border highlands each week of each month, from April through August: the brutal winds of spring, when gales off the desert gust above seventy miles an hour and the occasional snow squall turns my peak white; the dawning of summer in late May, when the wind abates and the aphids hatch and ladybugs emerge in great clouds from their hibernation; the fires of June, when dry lightning connects with the hills and mesas, sparking smokes that fill the air with the sweet smell of burning pine; the tremendous storms of July, when the radio antenna sizzles like bacon on a griddle and the lightning makes me flinch as if from the threat of a punch; and the blessed indolence of August, when the meadows bloom with wildflowers and the creeks run again, the rains having turned my world a dozen different shades of green. I've seen lunar eclipses and desert sandstorms and lightning that made my hair stand on end. I've seen fires burn so hot they made their own weather. I've watched deer and elk frolic in the meadow below me and pine trees explode in a blue ball of smoke. If there's a better job anywhere on the planet, I'd like to know what it is.

The work has changed remarkably little over the course of the past century, except in its increasing scarcity. Ninety percent of American lookout towers have been decommissioned, and only a few hundred of us remain, mostly in the West and Alaska. Nonetheless, when the last lookout tower is retired, our stories will live on. Jack Kerouac worked a summer on Desolation Peak in the North Cascades in 1956, an experience he mined for parts of two novels, *The Dharma Bums* and *Desolation Angels*. He secured the job through a recommendation by his friend the poet Gary Snyder, who worked summers on two different lookouts in the same national forest and wrote several fine poems about the experience. During the 1960s and '70s the old raconteur Edward Abbey worked as a lookout in various postings, from Glacier National Park to the Grand Canyon. He wrote two essays on the subject and made a fire lookout the main character in his novel *Black Sun*, the book he claimed he loved most among all his works. And Norman Maclean, in his great book *A River Runs Through It*, wrote a lightly fictionalized story about his one summer as a lookout on the Selway Forest in northern Idaho, over the Bitterroot Divide from his home in Missoula, Montana.

Based on their reminiscences, I'm pretty sure the qualifications to be a wilderness lookout remain the same as they ever were:

- Not blind, deaf, or mute—must be able to see fires, hear the radio, respond when called
- Capability for extreme patience while waiting for smokes
- One good arm to cut wood
- Two good legs for hiking to a remote post
- Ability to keep oneself amused
- Tolerance for living in proximity to rodents
- A touch of pyromania, though only of the nonparticipatory variety

Between five and fifteen times a year I'm the first to see a smoke, and on these occasions I use the one essential fire-spotting tool—aside, of course, from a sharp pair of eyes, augmented by fancy binoculars. That tool is the Osborne Firefinder, which consists of a topographic map encircled by a rotating metal ring equipped with a sighting device. The sighting device allows you to discern the directional bearing of the fire from your location. The directional bearing—called an azimuth—is expressed by degree markings along the outside edge of the ring, with 360 degrees being oriented with true north. Once you have an azimuth, you must then judge the fire's distance from your perch. The easiest way to do this: alert another lookout able to spot the smoke, have her take her own azimuth reading, and triangulate your lines. We lookouts call this "a cross," as in: *Can you give me a cross on this smoke I'm seeing at my azimuth of 170 degrees and 30 minutes?* If this can't be done—the smoke is too small to be seen by another lookout or its source is hidden by a ridge—you're thrown back on your knowledge of the country. Protocol dictates that you locate each fire by its legal description, or what we in the trade simply call its legal: ideally within one square mile, by township, range, and section, the square-ruled overlay on American property maps.

A new smoke often looks beautiful, a wisp of white like a feather, a single snag puffing a little finger of smoke in the air. You see it before it has a name. In fact, you are about to give it one, after you nail down its location and call it in to dispatch. We try to name the fires after a nearby landmark—a canyon, peak, or spring—but there is often a touch of poetic license involved. Some years there is more than one fire in a place called Drummond Canyon; knowing this, I'll name the first the Drum Fire, the second the World Fire, keeping in my hip pocket the possibility of a third fire I can call Drummond. Or say a fire pops up in Railroad Canyon, but there's already been a Railroad Fire that year. Something like Caboose Fire would be acceptable.

The life of the lookout, then, is a blend of monotony, geometry, and poetry, with healthy dollops of frivolity and sloth. It's a life that encourages thrift and self-sufficiency, intimacy with weather and wild creatures. We are paid to master the art of solitude, and we are about as free as working folk can be. To be solitary in such a place and such a way is not to be alone. Instead one feels a certain kind of dignity. There are other lookouts on other peaks in the same forest with stunning stretches of country beneath them—ten of them still in the Gila, to be precise—but none of them quite like mine. Dignity and singularity: these are among the blessings of solitude in a high place.

I harbor no illusion that our job will last forever. Some in the Forest Service have predicted our obsolescence within a decade, owing to more powerful radios, more sophisticated satellites, even drone airplanes—the never-ending dreams of the technofetishists. For now we remain indispensable because the Gila is so rugged that direct communication between crews and dispatchers can't be had from certain places. Lookouts are sometimes the only link to the outside world for backcountry crews of all kinds: fire, trails, Game and Fish. Just as important is the fact that we remain far less expensive than continuous aerial surveillance. Still, there's little doubt I'm practicing a vocation in its twilight.

DURING MY TIME IN THE GILA, I've been witness to one of the most remarkable ongoing revolutions in land management ever undertaken by the Forest Service. For millennia fires burned here with no effort to suppress them. Only with the coming of the Forest Service in the early 1900s did that change, as the government used all the tools and tactics of warfare on what it deemed a deadly enemy. Almost from its infancy the Forest Service hewed to an unyielding goal: to suppress every single fire as soon as it was detected. In 1934 this approach was codified in the so-called 10 a.m. policy, the aim being to have a fire contained by 10 a.m. on the morning after it was first spotted, or, failing that, by 10 a.m. the next morning, and so on. For another several decades that remained the protocol on every single wildfire in the West. By the late 1960s, ecologists and fire officials had begun to understand just how badly the total suppression strategy had warped whole ecosystems. A tinderbox of thick fuels had built up, primed to explode.

Here on the Gila, the ancient fire regime consisted mainly of cool-burning surface fires moving through grass and open pine savannas, fires that consumed ground fuels but preserved the mature forest canopy. These fires recurred, on average, once or twice a decade until about 1900. Then, amid heavy livestock grazing and an absence of fire, piñon and juniper crowded onto grasslands where quick-moving ground fire had once kept the saplings in check. (Most careful scientists are quick to claim that climate may also have played a role in piñon-juniper colonization.) Ponderosa took hold in dog-hair thickets in the middle elevations, where fires had previously burned over large areas every two to eight years. In the highest elevations, the effects of fire suppression are less stark, since the subalpine belt of spruce and fir typically burned in stand replacements every hundred to three hundred years. Still, it seems safe to surmise that an absence of both small and large crown fires crowded the aspen, which thrive in places disturbed by fire. And no one disputes that as the twentieth century drew to a close, fires were burning hotter over larger and larger areas of the West, a result of the militaristic ideology that for most of the past hundred years painted fire as the enemy of forests.

In the 1970s, a small group of firefighters on the Gila began, tentatively at first, to experiment with an approach that upended three quarters of a century of Forest Service dogma. Having studied the effects of the suppression regime, they wondered whether it might benefit the health of the forest to return fire to the land. Initially called "prescribed natural fire," this policy involved letting some fires burn unchecked toward the end of fire season, deep inside the wilderness, after the onset of rain had reduced the danger of a big blowup. The results were promising. The forest showed incredible resilience, green shoots of grass poking through the black within weeks or even days of a burn. Springs that hadn't run in years began to flow again, no longer tapped out by unnaturally crowded vegetation. Emboldened by these experiments, fire managers began to allow larger burns earlier in the season, trying their experiments in various fuel types from semiarid grasslands on up to ponderosa savanna and even mixed conifer.

It would be difficult to exaggerate how radical an experiment those first "prescribed natural fires" were within the prevailing mind-set of the Forest Service. But their success caused others to take note. In 1978, based on early results from the Gila and the Selway-Bitterroot in the northern Rockies—the two national forests that pioneered the let-burn approach—the Forest Service officially rescinded the 10 a.m. policy.

What's happened here is akin to species reintroduction. Think of fire as an endangered species, which, after its nearly century-long absence, we're attempting to restore to a land that evolved with its imprimatur. The two wilderness areas at the heart of the Gila get hit by lightning, on average, 30,000 times per year, and they comprise less than a quarter of the forest's total area of 3.3 million acres. In the United States, only the Gulf coast of Florida surpasses the Southwest in density of lightning strikes, and there it comes with far heavier rains. Combine the frequency of lightning with an arid climate and one begins to understand why this is the most fire-prone landscape in America. In my eight years on the peak the forest has averaged more than 200 wildfires per season— a low number by twentieth-century norms. Each year now, several lightning-caused fires on the Gila are allowed to burn for weeks or even months at a time, sometimes over tens of thousands of acres. Often they remain benign surface fires. They chew up the ground-level fuel that's built up over the decades. Occasionally they torch in stands of mature forest, establishing a mosaic of open meadows. Even the hottest fires, with their apocalyptic visions of black smoke pluming thousands of feet in the air, have their place. As devastating as they appear, they constitute the birthday of aspen and oak, which seize their turn in the succession of forest types natural to this part of the world, a succession determined in large part by fire. It's not as if anyone can say with certainty which fires should be put out and which allowed to burn. Firefighters still suppress more than 90 percent of all wildfire starts here. The experiment is ongoing, as much art as science, and I feel fortunate to have held a front-row seat for the show.

The lookout's purpose has evolved along with our understanding of fire's intrinsic value. No longer are we merely a first-alarm system alerting authorities to the presence of a disruptive force, whereupon men and women descend in parachutes and helicopters to quash it. Now, on some smokes, we work with fire managers and "fire-use modules," keeping radio contact with crews on the ground as a fire does its ancient work, offering eyes in the sky as we monitor the course of a blaze that may, in the end, spread across dozens or even hundreds of square miles. In 2003, for instance, I watched a fire in the heart of the Gila Wilderness burn for two months over 98,000 acres, an area two and a half times that of the District of Columbia. Such fires, long presumed to be malignant intruders in a fragile landscape, are now welcomed, even encouraged—at least here. Fires on the edges of Salt Lake City and Boise, where deference to life and property remains paramount, are another matter.

None of these experiments with wildfire would have been possible if the landscape hadn't been preserved in something approaching its primeval state. From my peak I have the good fortune to look out on a land of great significance in the history of American ecology: the Gila Wilderness, the first stretch of country in the world to be consciously protected from incursion by industrial machines. That designation was largely the work of one of the most important figures in early-twentieth-century America, the writer and conservationist Aldo Leopold. Many scholars have called *A Sand County Almanac*, his posthumous book of essays, the bible of American environmentalism; Leopold is often referred to as a prophet. Like most who attract such labels, he honed a late message that was both radical and difficult, even while expressed with an aphoristic beauty seldom matched in American writing on the natural world. To arrive at such a style, not to mention its moral depth, took Leopold a lifetime of close observation, deep thinking, and a willingness to change that thinking when his ideas proved an insufficient match for the land's complexity. Though he's often associated with Wisconsin, where he lived the last twenty years of his life, it was here in the Southwest—in the country stretching from the Rio Chama of northern New Mexico south to the Gila and west across Arizona to the Grand Canyon—that Leopold unlearned most of what he assumed and had been taught about man's relationship with the earth. In this vast and arid terrain encompassing four of the six life zones, he developed an influential argument in favor of wilderness with profound effects on the American landscape, some of them felt most tangibly on the stretch of country outside my windows.

In 1924, as an assistant district forester in New Mexico and Arizona, Leopold drew a line on the map encompassing four mountain ranges and the headwaters of the Gila River, a line beyond which nothing motorized or mechanized would be allowed to travel. In 1980 another roadless area was preserved alongside it, this one named in honor of Leopold: more than 200,000 acres running along the crest of the Black Range and down its slopes to the east and west, into the foothills and mesas in its shadow. This would seem an appropriate homage to a man who not only conceived of and drew the boundaries of the world's first wilderness, but founded the field of wildlife management, helped pioneer the study of tree rings for evidence of fire history, and articulated a philosophy—the "land ethic"—that inspired subsequent generations of environmental thinkers.

The Gila Wilderness served as a guiding example for that capstone of the American preservation movement, the 1964 Wilderness Act. By leaving a big stretch of country unroaded and unpopulated, it also removed the major obstacles to large-scale burns: the presence of private property and the existence of communities impacted by smoke. For we must acknowledge a simple fact: smoke and flames make people nervous, whether they're watching from the front porch of a mountain cabin or from the back deck of a home in town. The two wilderness areas at the heart of the Gila give forest officials a buffer of safety between humans and fire. In many places throughout the West, no such buffer exists between wildlands and the "urban interface," to use the jargon of our day. It is from those places where most Americans get their images of wildfire—the drama of the fight, the tragedy of the charred home. This book, I hope, will offer another view of fire and its place in nature, a view too little glimpsed on our television screens.

What follows, then, is more than just a personal drama—though living alone on a high mountain, in the company of abundant wild creatures, surrounded by a landscape prone to burn, does provide an impressive stage set for a drama of the self. I often think that if there's such a thing as the oldest story on earth, it is a story of fire, the marriage of fuel and spark. Despite all the vitriol we've directed at it, despite all the technology we've deployed to fight it, wildfire still erupts in the union of earth and sky, in the form of a lightning strike to a tree, and there is nothing we can do to preempt it. The best we can do, in a place like the Gila, is have a human stationed in a high place to cry out the news. If this gets to sounding borderline mystical, as if I've joined the cult of the pyromaniacal, all I can say is: guilty as charged.

1

————— ※ —————

APRIL

————— ※ —————

Oil the saws, sharpen axes,
Learn the names of all the peaks you see and which is highest—
    there are hundreds—
Learn by heart the drainages between
Go find a shallow pool of snowmelt on a good day, bathe
    in the lukewarm water.

    —Gary Snyder, "Things to Do Around a Lookout"

*Into the Black Range ✻ thwarted by snow
& saved by snow ✻ a view from on high
✻ unsettled by solitude, troubled by wind ✻
some walks with the dog & bears we have
seen ✻ cutting wood the old-fashioned way ✻
the symbiosis of grass & fire ✻ a visit from
the mule packers ✻ smoke in Thief Gulch*

APPROACHED FROM THE VALLEY of the Rio Grande, the Black Range is only modestly impressive, a low dark wall of irregular height seventy miles long, rising above tan foothills. All around are mountains more exposed in their geology, more yielding of their ancient secrets. The muted shapes and dark color of the Black Range give it an appearance of one-dimensionality, perhaps even unreality, as if it were a painted backdrop in a low-budget Western. These are not picture-postcard peaks, serrating the sky with shark-tooth shapes of bare rock. Instead they're a doubtful chimera on the edge of the desert, a sky island seeming to shimmer in an April haze.

The Black Range was once a part of Apache country, one of the major reasons it was slow in coming under the domain of the American government. Geronimo was born to the west near the forks of the Gila River; his fellow chief Victorio, of the Warm Springs Apache, made his home just to the north, at Ojo Caliente, though he knew the Black Range intimately, having used it as a hunting ground and refuge from the summer heat. He and his Chihenne followers fought the U.S. cavalry here as late as 1880, and a number of Buffalo Soldiers and their Navajo scouts did not have the luck to leave the Black Range alive. Their graves can still be found if you know where to look.

For me, the first leg of the trip to the crest is simple and comfortable: a climate-controlled pickup truck, Satchmo on the stereo, my dog Alice on the seat beside me. We glide across the bed of an ancient inland sea, which locals call the flats. Beyond the little town of Gaylord the road curves to contour with Trout Creek, a stream with sources high in the mountains, denuded in the lower elevations by decades of overgrazing. At Embree, population three dozen, the road leaves the creek to begin its climb through the foothills. For fifteen miles there's not a straightaway long enough to allow me to pass a slow vehicle. Locals, if they see me come up behind them, will pull into one of the gravel turnouts and offer a wave as I pass, but if I find myself behind tourists, I will dial down my speed and practice the virtues of charity and patience. It is a magnificent drive, and I can hardly fault them for taking it leisurely.

I can't help but hurry, despite the sweeping views. Where I'm headed the views are a whole lot better, and besides I've been gone too long. Seven months of hustle in the world below provide more than sufficient acquaintance with the charms of my winter career. But that is behind me now—or rather, I should say, beneath me.

16

Norman Maclean once wrote that "when you work outside of a town for a couple of months you get feeling a lot better than the town and very hostile toward it." I felt hostile and superior before I even left. Tending bar will do that to a man, although it also allows him to leave on short notice, and in so doing hurt no one's feelings but those of the regulars who've come to depend on his ear.

More than one winter has found me working in a Silver City dive that beckons to the thirsty with a classic neon sign of a cactus in the foreground and a horseman drifting alone into the distance. My most reliable customer was an Oklahoma hillbilly with a Santa Claus beard, whose wit and wisdom is best exemplified by a statement I've heard more than once from his beer-foamed lips: "Thing about them Aye-rabbs, they breed faster'n we can shoot 'em. Kinda like them Kennedys." Weekend entertainment brought the biggest crowds and the best money, courtesy of heavy-metal bands with evocative names such as Dirtnap, Bowels Out, and New Mexican Erection. It all makes for an interesting counterpoint to summers spent alone far from town, but I'm tired of playing the role of enabler-priest in an unholy chapel.

At Wright's Saddle my drive is over, though the real pleasures of the journey have just begun. In a supply shack cluttered with helicopter sling nets and cases of military-style MREs (Meals, Ready-to-Eat), I leave eight boxes to be packed in later by mule, each box marked with its weight, to help the packers balance their animals. The boxes contain books, dry and canned food, dog food, two cases of double-A batteries, a Frisbee, a mop head, a bow saw, an ax bit. I double-check my own pack for all the immediate necessities: maps, binoculars, handheld VHF radio, freeze-dried food, my typewriter, some magazines, some whisky. Certain I've left nothing vital behind, I begin the final stretch of what has to be one of the sweetest commutes enjoyed by any hardworking

American anywhere. Alice leaps about, wagging her question-mark tail. She feels the same way I do.

Five and a half miles await me, five and a half miles of toil and sweat, nearly every inch of it uphill, with fifty pounds of supplies on my back. I can feel right off that winter has again made me soft. My gluteal muscles burn. My knees creak. The shoulder straps on my pack appear to want to reshape the curve of my collarbones. The dog shares none of my hardships. She races to and fro off the trail, sniffing the earth like a pig in search of truffles, while from the arches of my beleaguered feet to the bulging disk in my neck—an old dishwashing injury, the repetitive stress of bending forward with highball glasses by the hundred—I hurt. Not many people I know have to work this hard to get to work, yet I can honestly say I love the hike, every step of it. The pain is a toll I willingly pay on my way to the top, for here, amid these mountains, I restore myself and lose myself, knit together my ego and then surrender it, detach myself from the mass of humanity so I may learn to love them again, all while coexisting with creatures whose kind have lived here for millennia.

Despite human efforts to the contrary, it remains pretty wild out here.

Along the path to the peak the trail curves atop the crest of the Black Range first to the west, then back to the east, always heading eventually north. Despite the wild character of the country, there is evidence of the human hand all along the way, not least in the trail itself, an artificial line cut through standing timber. The wilderness boundary too speaks of a human imprint: a metal sign nailed to a tree suggests you leave your motorized toys behind. Halfway to the top the trail passes an exquisite rock wall, handiwork of the Civilian Conservation Corps of the 1930s, when the New Deal put thousands of Americans to work on the public lands of the West. The wall holds the line against a talus slope above it, keeping the loose rock from swamping the trail. Seventy years later the wall is as solid as the day it was built, as is the lookout tower where I'm headed, another CCC project that replaced the original wooden tower built in the 1920s.

For a while thereafter the trail follows an old barbed-wire fence, a relic of a time, not that long ago, when cattle grazed these hills. High on the trunks of old firs and pines hang a few white ceramic insulators, which once carried No. 9 telephone line down from the lookout. Having spent something like a thousand days in this wilderness over the past decade, I've noticed all of these features of the hike many times. And yet there are always surprises: a tree shattered by lightning, a glimpse of a black bear, the presence, in a twist of mountain lion scat, of a tiny mammalian jawbone—evidence of the dance of predator and prey.

The surprise this time arrives a half mile below the peak: a stretch of hip-deep snow. It swallows the trail amid an aspen grove on the north slope, and there is no shortcut from here, nothing to do but slog on through. For a few steps I'm fine. The crust holds. Then it collapses beneath me, and I posthole to about midthigh. I try to lift one leg, then the other, but I feel as if I'm stuck in quicksand. I'm not going anywhere unless I lose my pack.

With its weight off my back I can extricate myself, but there remains the problem of how I'm going to get both it and myself through the next 800 yards. Upright on two legs, 220 pounds of flesh and supplies on a vertical axis, I will continue to sink and struggle. The snow is wet and granular, melting fast; in just a few weeks it will be gone. But not yet. There is nothing to do but become a four-legged creature, distribute my weight and the weight of my pack horizontally, and crawl. Alice looks at me as if I've lost my marbles—she even barks twice, sensing we're about to play some kind of game—but now when I punch through the surface crust I don't plunge as deep, and with the aid of my arms I can drag myself along, bit by bit, crablike up the slope. Alice runs ahead, returns, licks my face, chews a hunk of snow, bolts away again—she's delighted by my devolution to four-legged creature, though I can't say I feel likewise. All I can see from this vantage is an endless field of white broken only by tree trunks, and I possess neither her agility nor her lightness of foot.

After a twenty-minute crawl I reach the clearing at the top. Normally I would rejoice in this moment—home, home at last, mind and body reunited on the top of the world—but my hands are raw and red and clublike from the cold, and my pants are soaked from crawling in snow. The sun is dropping fast and with it the temperature. If I don't change clothes and warm up, I'm in trouble.

The cabin is filthy with rat shit and desiccated deer mice stuck to the floor, dead moths by the hundreds beneath the windowsills, but these are problems for later. I start a fire using kindling gathered late last year with precisely this moment in mind. I strip off my pants and hop into dry ones, never straying far from the pot-bellied stove.

Once I'm warmed through, I tend to the next necessity: water. Just outside the cabin is an underground tank, 500 gallons of rainwater captured by the cabin's roof and funneled through a charcoal filter into a surplus guided-missile container. It's the sweetest water I've ever tasted, despite what holds it, and to keep it that way I lock the ground-level lid for the off-season. This winter, I discover, someone tried to get at it and broke off a key in the lock. Why a visitor thought he could open a U.S. Forest Service padlock—it is stamped *USFS*, unmistakably—is difficult to fathom, but I've spent enough time here to know that people do strange things alone above 10,000 feet. Maybe it's simple lack of oxygen to the brain.

My water supply shall remain, for now, inaccessible. The snow I so recently cursed is my savior. Melted in a pan on the wood-stove, strained of bits of bark and pine needle, it tastes nearly as sweet as I remember the cistern water, with just a tincture of mineral earth. My thirst quenched and my hands warm, I heat some snowmelt and freeze-dried minestrone to head off the roiling in my stomach.

The sun drops over the edge of the world. The wind comes up, gusts to near forty. I jam the stove with wood, unroll my sleeping bag on the mattress in the corner, and free-fall into untroubled sleep.

FIVE HOURS LATER I WAKE to find Alice has joined me in the bed. I can't say I mind the added warmth of her next to me. It is still only 2 a.m., but the stove has burned out. I revive the fire from the ashes of itself, drink some more snow water. Outside the wind screams in the night, gusting now to fifty, buffeting the cabin like some rude beast up from the desert. I pull on an extra pair of wool socks, a down vest, a stocking cap, gloves. Time for a look around.

Some of my fellow lookouts live in their towers, spacious rooms with catwalks around the exterior. My tower is small and spare, seven-by-seven, purely utilitarian—more office than home. It can hold four people standing, assuming they're not claustrophobic. At fifty-five feet tall, it is one of the highest lookouts still staffed in the Gila. It had to be built high to offer sight lines over the trees—my mountaintop being relatively flat—and in my more poetical moods I think of it as my mountain minaret, where I call myself to secular prayer.

Near the top the wind grows fiercer. I grasp the handrail, climb the last flight of stairs, shoulder my way through the trap-door in the floor, and there it is: my domain for the next five months, a stretch of earth cloaked in the mystery of dark, where the Chihuahuan and Sonoran deserts overlap and give out, where mountains 35 million years old bridge the gap between the southern Rockies and the Sierra Madre Occidental, and where on the plains to my north the high-desert influence of the Great Basin can be felt. I am perched on the curving southern spine of the Black Range, with a view all the way up its east side. The range runs almost straight north and south and for much of its length marks the Continental Divide. The waters on the east flow to the Rio Grande and eventually on to the Gulf of Mexico, while the waters to the west join the Gila River on its 650-mile journey to the Colorado at Yuma, Arizona—a journey it rarely completes, thanks to thirsty Arizonans.

Even to drive the circumference of country I can see from here by daylight would take two full days—plus an international border crossing. It is a world of astonishing diversity and no small human presence. To the east the lights of Truth or Consequences twinkle like radioactive dust; in the southeast, just beyond the horizon, El Paso and Juarez glow like the rising of a midnight sun. Along an arc to my south are the lights of other, far smaller towns, lights that remind me I sit at the juncture of more than one transition zone, not just the meeting of different biomes but the wildland-urban interface.

The town lights are quaint and even rather beautiful at this remove, but the wildlands are what draw my eye whenever I climb the tower at night. In a quadrant from due west to due north there is no evidence of human presence, not one light to be seen—a million uninhabited acres. A line running northwest of where I sit would not cross another human dwelling for nearly a hundred miles, a thought that never fails to move me.

I say move me, but that doesn't quite do justice to the feeling.

In fact, if I'm to be honest about it, on this my first night back in the tower I find myself hopping around like some juiced-up Beat poet, but instead of shouting Zen poetry and gentle nonsense I start hollering profanities, turning this way and that trying to take it all in but it's just so huge there ain't no way. How can I explain this outburst, other than to say no disrespect to the faithful intended? I grew up Catholic, after all. Curse words and states of intense feeling have always seemed to me a natural match.

Satisfied with the extent of my elbow room, I drop back through the trapdoor, shimmy down the steps, hurry back to the warmth of the cabin. I feed the potbellied stove with pine I split last August. For an hour or more I lie awake in the dark, listening to the howl of the wind in the trees, resisting an urge to call the dog back to the bed and thereby spoil her beyond reckoning.

IT'S A LOT OF WORK setting up to be lazy.

For seven months I surrender the cabin to the creatures, if they can get in. This winter they've been aided by a citizen user of the national lands who decided to break a window in the side door, for reasons unknown and unknowable. (Five miles afoot seems a long trip to commit a petty act of vandalism.) Every surface is covered in dust from the winter winds working their way through the crevices, blowing in through the window. Not even the dishes in the cupboards are spared a fine grit. All must be washed; I heat some water in a basin for just this purpose. While I wait for it to warm I cover the window with plastic sheeting and duct tape, measure it for a Plexiglas replacement to be hauled up later.

Next all evidence of the rodents must be expunged, especially the smell. I pry the dead deer mice off the floor and throw them in the fire. Dustpans full of moths meet the same fate. The pack

rats have made nests in the bedroom cabinets, filthy conglomerations of pine needles, plastic spoons, Band-Aids, pages torn
from magazines, random playing cards. These too I burn. The
rats have the unfortunate habit of urinating in the corners and
defecating where they sleep, filling the cabin's atmosphere with a
bitter, ammoniac stench—redolent of certain New York streets I
have known. To ward off hantavirus I spray the floor with a solution of water and bleach, then mop with water and pine-scented
soap, and finally mop once more with just water.

Once the place is somewhat livable I figure I'd better test my
radio, let the below world in on the fact that I've arrived here
safely. Not that anyone's all that concerned. Once we're sent up
our hills, we lookouts are largely forgotten, which is just the way we
like it. Until we see some lightning, we tend to follow the model of
Victorian-era children: do not speak unless spoken to. An exception is this courtesy call to let the office manager know I'm alive
and that my tower withstood another winter of battering wind. I
climb a ladder and affix a magnetized antenna to the cabin's roof.
I plug a handheld mic running from the antenna into the side of
my Bendix-King radio. A red light blinks on when I press transmit.

"Black Range District, Apache Peak."

"Go ahead Apache Peak."

"Just letting you know I'm on the peak safely. All's well here."

"Copy. Anything you need?"

"Affirmative. Some joker broke off a key in the lock to the
cistern. I'll need a bolt cutter for the old lock and a new lock to
replace it."

"Copy that. We'll send those up with the packers."

"Tell them to wait two weeks or they won't get their animals
through the snow on the north slope."

"I'll let 'em know. You stay warm up there if you can."

"Copy. Apache clear."

I rather like the laconic style of these conversations, which results, I often feel, from a kind of mutual incomprehension. The person on the other end can't fathom why I'd spend most of the summer alone in a little tower, far from running water, cold beer, and satellite TV. I cannot conceive of why you'd join the Forest Service only to spend your days shuffling paper under banks of fluorescent light. To him I must seem a little crazy, even something of a masochist, a perception I do nothing to dispel on those rare occasions—a couple of times a year, to pick up my radio and turn it back in—when I visit the office.

Some years ago, when forced to undergo a "training and orientation" session prior to the beginning of fire season, I raised my hand in protest at having to sit through a seminar on sexual harassment in the workplace. The only conceivable way this affects me, I pointed out, is in my choice of whether to harass myself with my left hand or my right. This heresy was greeted with a gimlet eye and a stern word—not to mention a few muffled chuckles—but it had the desired effect. I was not invited to attend such a session again. The less normal I can make myself appear, the less likely I am to be drafted into some sort of nonlookout work, or any task involving contact with the public in an official government capacity. I must act the part well enough—a goofy hermit with a weird beard and a faraway look in his eyes—because for most of my season I am left utterly and blissfully alone. My radio contact is where he wants to be, I'm where I want to be, and in this fashion we preserve a bit of the diversity of our human experience. Not unlike the vigorous diversity of the forest out my window, where the ponderosa at their healthiest grow in open parklands, the trees spaced fifty or a hundred feet apart, while the aspen cluster in dense groves with a shared root system.

I am not unfamiliar with his world. In fact, I once inhabited a more extreme version of it: four years at a desk, in front of a computer, enclosed by the steel and glass colossi of lower Manhattan. He at least has a view of mountains out his window, far off in the distance, which trumps the view I once had of Jersey City. I found rather quickly during my peonage in newspapers that I did not have the requisite temperament for such work—the subservience to institutional norms—and that I gained very little in the way of psychic ballast from the attachment of my name to that of a well-known brand in American journalism. Maybe it was a case of egocentricity, but I discovered I had things to say that could not be said in the pages of a daily newspaper. Plus, when the weather cooperates, I prefer to work shirtless.

Let there be no illusion about the purity of my solitude. Not only do I see the occasional hiker, but I am here for ten days followed by four days off. This routine will hold for the next four and a half months, give or take a couple of weeks, depending on the fire danger rating, the Forest Service budget, and the timing of the summer rains. My pilgrimage to town every other weekend allows me a restorative dose of civilized pleasures. I rendezvous with my wife, Martha, take a hot shower, drink a cold beer. I catch up with friends and shoot some pool to keep my game from going entirely to pieces. By about the third day, I'm ready to be back in the woods, and come midseason I will often forgo time spent in town and instead—with Martha's indulgence, perhaps even her company—grab my fly rod and head for trout streams reachable in no less than half a day's walk.

In most ways I have it easier than my predecessors. Once the lookout lived in an earth dugout and hauled water from a spring a half mile away. When he spotted a fire, he was expected to saddle a horse and ride straight for the smoke, tools at the ready to put the thing out. There was no such thing as a day off. The lookout came in the beginning of the season and stayed until the end. My days off are covered by a relief lookout. I have four sturdy walls and eight windows, a refrigerator that runs on propane, a mattress flown in years ago by helicopter. The entirety of my duties are five: report the weather each morning, answer the radio, relay messages when asked, call in smokes when they show, and keep an eye on fire behavior with the safety of crews in mind. This time of year, in the absence of lightning, a smoke is unlikely. If one appears, it will have been caused by humans—an abandoned campfire, a cigarette tossed from a car—and will appear near a road. The open meadow to my southwest affords me a view, without my climbing the tower, of a good chunk of the roaded country for which I'm responsible; a quick glimpse from the tower twice an hour suffices to cover the country to the southeast, south, and west. To the north there are no roads.

In these early days, as I work on sprucing up the place, I fulfill the command embedded in my job title and "look out" maybe two hours a day—fifteen minutes every hour. That will change with an increase in fire danger, but for now the majority of my time is spent cleaning, cutting wood, sleeping, reading, and cooking, a regime I follow with unyielding discipline. Alice sits in the meadow and gnaws on a mule deer antler she found; she loves indolence as much as her master and appears to feel no pangs of conscience about doing merely what pleases her. More than one visitor has remarked that she must be the luckiest dog in the world. I can't speak to the experience of all dogs everywhere, but it's true that Alice finds a great deal of joy here. From the moment my backpack emerges from the closet at home, she begins to pace and pant in excitement, her tail working in figure eights, for she knows what this contraption signals: a trip into the woods.

When Martha and I first laid eyes on her, she was an underfed, eight-month-old puppy, locked up in a cage at the dog pound. She was all black except for a white stripe running from her throat to her chest, a coloring that made her appear as if she were always dressed for a black-tie affair. When we paused to look in on her, she rose from her grimy and pathetic little bed and gently licked the hands we proffered to pet her. From then on it was only a matter of formalities—paperwork and vaccinations and the payment of a fee for having her spayed. Once we'd been licked, everything clicked.

There were the usual difficulties at first. She'd been on the streets of our town for unknown months, a scavenger's existence, and backed away in skittishness whenever a hand was inadvertently raised above her head. More than once we woke in the morning to find a brown pile or a yellow puddle in the middle of the kitchen floor. When we were gone a portion of the day, she greeted our return by jumping at us with her paws extended, as if she wished to dance; often this resulted in livid scratches on our forearms, marks of her overzealous need for love.

Still, she made steady progress on the path to domestication. She romped agreeably with other dogs in the park, learned to heel and fetch a stick, brought her bodily functions under control. We taught her the other doggy basics: to sit and shake, lie down and roll over, jump and grab a stick in her teeth. She learned what was meant by the words, *Are you hungry?*, and her answer, expressed in upraised ears and a quick pirouette toward her food dish, was never not *yes*.

Nine months into our acquaintance, she learned she had two homes, not one. A summer getaway in the hills. She took to the place instantly, and her change in personality was striking. Whereas in town she'd been a timid creature, afraid to step into the yard and do her business when the wind was up, on the peak she'd lie in the meadow all afternoon in a forty-mile-an-hour gale, nestled on the leeward side of the helicopter landing pad. She recovered a streak of independence, wandering off by herself in search of bones, antlers, small mammals to startle and chase. Mornings when deer came to the meadow she'd run them off. She guarded her territory with a nervy vigilance undermined only by a high-pitched yip almost comical in its lack of menace. Hikers were often startled to hear their approach announced in advance.

I found it amusing and even a little exhilarating to watch her regain a touch of her former wildness, that wildness which lies dormant in the hearts of most domesticated creatures. The way she attacked a mule deer antler with ferocious jaws, the way she marked her presence by dribbling a few drops of urine on a bob-cat turd—these spoke of atavistic urges, instincts toward survival and even dominance in a world of struggle and strife. Back in town on my days off, she resumed her sedentary existence, nap-ping on the couch all afternoon, begging for the chance to snuggle between her two softhearted masters. The woods invigorated her, and she intuited the difference between that world and her other, modulating her behavior accordingly.

On the mountain she proves an agreeable companion in more ways than one. She never insists on conversation. She requires only a dish of water, three squares a day, and a scratch now and then behind the ears— the perfect partner in solitude. When I sign off the radio and descend the tower at 6 p.m., she's immediately at my side, urging me to take her for a hike. Without her here I might be tempted to bag my evening stroll, go straight to dinner and whisky and a ball game on the radio, but her enthusiasm always wins the day, and I'm the better for it.

So is she, for she loves to run, loves the pursuit. If you walk in wild country five miles a night you're bound to scare up a creature or three. In her time she's flushed elk, mule deer, wild turkey, quail, grouse, skunks (unhappily), rabbits, squirrels, chipmunks, rats, and mice. More than once we've encountered a bear. The first time, she smelled it before either of us saw it; her nose to the ground, her back up like a razorback hog's, she snorted and growled and offered a lower, more menacing bark than I'd heard from her before. I spoke to her in a soothing tone, using one of the dozen nicknames we've given her over the years.

It's okay little spooky one, little Spookeen. Stick close and we'll be just fine, you and me, Spooky Groo. Yes, yes. Atsa girl.

Every nerve ending in my body needled my skin as we followed the switchbacks down the northeast slope of the peak, through lovely fir forest and down into aspen, with Alice's ears upraised just a touch, her tail moving back and forth in the shape of a machete. All of a sudden something crashed like a tree on the far side of a little open meadow—except it kept on crashing after itself, ever more softly as it went, like a domino arrangement of trees, each one smaller than the next. Before it was gone I caught a blurred glimpse of its cinnamon back and black rump.

Alice took off like a shot, yipping her funny yip, racing toward the meadow, and I called after her, *Hey hey hey hey hey hey hey*, and she put on the brakes, skidded in the needle-cast, and responded, twice, *Ruff . . . . . uhhrrruff.*

Good girl, I said. *Good Spookeen.*

She held her ground. After a few more gruffs and growls, we kept on our way.

Just as she smells things I've yet to see and never may, so I sometimes see things she has yet to smell. Her nose supplements my eyes; my eyes sometimes preempt her nose. One evening we took the faint path north of the peak, out along the Ghost Divide, a series of knolls and ridges leading into some of the wildest country in the Aldo. Locust and Gambel oak crowded the trail so that every once in a while it disappeared entirely, discernible only by the blazes on the trees. We made it nearly three miles, just above the point where we needed to turn around to make it back to the top by dark, when ahead on the trail I spotted something that looked like a charred stump, except that it was moving and had a tan-colored snout. I turned and looked for Alice, who for once trailed behind me; I softly called her away from her hunt for ground birds. When she pulled up next to me I grabbed her collar, turned her around, and led her back in the direction we'd come, not letting go until I was sure she was with me and understood the seriousness of our intent to retrace our steps. (She much prefers the loop hike to returning over ground she's already scoured.) By some miracle she neither saw nor smelled the bear; perhaps just as lucky, the bear did not appear to see or smell us.

THERE ARE TIMES, EARLY EACH season, when the solitude unsettles me. I have spent my winter kibbitzing with drinkers, after all, a variety of human not known for reticence. Night after night of the usual story: problems with women, problems with men, problems with money, problems with kin. A man grows accustomed to nodding his head in feigned sympathy and reaching for the bottle, two gestures that can only in the end corrupt, especially when performed in unison and for money. Following the drift and weave of my own thoughts is no simple task at first. Like any refined art it begs practice, and I am for the moment out of step.

The wind doesn't help. Two days are all it takes for the noise to insinuate itself in my cranium, a fluctuating symphony of sound: whisper and whistle, moan and roar. It gives a texture to the days, gaining power through the afternoon, barreling through the night, easing briefly at dawn. One morning it ceases altogether, and the sudden quiet knocks me out of balance, forces me to recalibrate the terms of the truce I've made with solitude. At sunrise I step gingerly into the meadow to piss, and all the world is still; I do not have to take care with my aim for fear of dribbling all over myself. I wonder: Does a snake feel pain when it molts a skin? My shedding of the social world is not without its element of discomfort. Pissing on the naked earth seems, obscurely, to help. So does singing remembered songs by Sam Cooke, Uncle Tupelo, and Hank Williams Jr. On the first calm night I build a bonfire in the stone circle and pace off my unease in the dark, howl a little at the moon, my bottle of whisky perched on a stump. The moon rises out of the desert, huge and orange and leering like a lopsided jack-o'-lantern. Flame and spirits and a weird pumpkin god of the sky: enough, together, to see me through the long cold night.

In the morning, reborn once more, I walk. My options are nearly limitless. Official trails and game trails fork in all directions. Every which way is down, down into variegated forest, trees dead on the ground, dead trees standing shorn of bark, trees corkscrewed by lightning, trees stunted by wind, trees with cankers and knobs, trees clawed by bears, trees grown together from two to make one, giant old fir trees five feet in diameter, aspen groves so thick you can't walk through them. On this day I choose the path to the spring. On the north side of the peak I hear the strange descending call of the Montezuma quail, otherwise known as the Harlequin quail, for the spiral black-and-white swirl on the side of the male's head. A queer-looking bird, it often freezes when a human draws near, as if pretending to be a piece of statuary; a person can come within a few feet of it before it rockets into flight. Nearer the spring a turkey gobbles, and I spot two hens lurching away through the trees. Amid a mature aspen grove, just above the spring, I hear the first hummingbird of the season, a broadtail. Time to make some sugar water and spy on the little creatures while they drink.

Time, as well, to cut wood. I figure I have enough to stay comfortable for seven more days, longer if I get good solar heating through the windows of the cabin in the afternoons, but it's always wise to work ahead. Rain could spoil the cutting for days. Overnight the mercury still drops into the teens. Being situated just inside the wilderness boundary—defined by the Wilderness Act as a place where motorized and mechanized machines are forbidden, "and man himself is but a visitor"—I am forced to use an axe and a handsaw. I've come to enjoy the simple rhythms of my arm moving a blade of metal teeth through pine and fir, and later on the thought of my labor will make me frugal with the resource of warmth. Some years a trail crew will fell a dead tree outside the wilderness boundary just to my south, then buck up the trunk into rounds I can split with an axe. Otherwise I cut mostly downed limbs of six inches' diameter or less, the size of the feeder hole in my stove. In my years here I've picked the ground clean a hundred yards in all directions, so the hunt becomes the greater part of the work—the hunt and the haul to the sawhorse.

LOOKOUTS ARE THE MOST PUNCTUAL people I've ever known. Each morning we sign on the radio at 8:00 or 9:00 a.m., depending on our set schedule, and because our transmissions are made over a forest-wide frequency, we cannot afford to lose track of time, unless we want to embarrass ourselves to unseen colleagues. Luckily, one lookout's announcement is a kind of wake-up call for all of us. The beauty of it is we could be naked, bleary-eyed, uncaffeinated, just rolling out of bed—and no one would know, as long as we could speak clearly over the radio the name of our mountain and the words "in service," at exactly the hour prescribed.

At 9:15 we recite our weather observations to the dispatcher: wet and dry temperatures, percentage of relative humidity, percentage of cloud cover, wind speed and direction, any lightning or precipitation over the past twenty-four hours. We measure humidity with a nifty tool called a sling psychrometer. It holds two thermometers, side by side, in a metal casing on the end of a chain. One of the thermometers has a small cloth sleeve over the bulb, which is dipped in water. Spun by the chain in the shade of a tree, the thermometers offer two different air temperatures, one wet, one dry. Using a chart calibrated by elevation, you can gauge the relative humidity by the difference between the two readings. On those rare occasions when the readings are the same, you have 100 percent humidity, which you'll have known already since you'll have been twirling your psychrometer in the rain.

One morning in late April we receive the unofficial signal that fire season has truly begun. Around 10:30 a.m. the dispatcher advises all personnel to stand by for the "fire-weather forecast." The man who recites it sounds as nervous as a grade-schooler chosen to read, in front of the class, a story with a few unfamiliar words. The peculiar thing is, only the numbers change. The words remain the same: high pressure, low pressure, relative humidity, upper- and lower-elevation temps, twenty-foot winds, transport winds, mixing height, ventilation category, Haines index, lightning-activity level . . . It's a morning poem of meteorology, and he never fails to bungle it in the beginning, mispronouncing a few words, audibly clearing the frog in his throat. Eventually he settles in and finds a rhythm, his self-consciousness ebbs, and the language of fire weather begins to sound like a Gary Snyder riff. The delivery is neither sonorous nor sensitive, but he can read, goddammit—he is predicting atmospheric conditions, he knows the whole forest is hanging on his every word, waiting for the day when the lightning-activity-level number rises from a one (no lightning expected) to a three or even four—but day to day he will never become a more fluid reader. He always falters out of the gate and slowly gains confidence and ends with his dignity repaired, and we are thankful for our poem.

And then the holy silence.

Time shapes itself around me in that silence, shape-shifts from mistress to shade, caressing and haunting by turn. Days pass in which there is nothing but wind, bending the pines to postures of worship of an unseen god in the east. The sun bores through the glass windows of the tower, solar heating at its essence. The world becomes the evolution of light. The almost imperceptible shift of color in the sky before dawn, the turn from midnight blue to sapphire. The way the mountains move through shades of green and blue and on through purple and black in the evening. The dark blue reefs of cloud in a backlit sky at twilight. A crimson lip at the edge of the world where the sun has gone, like a smear of blood, reappearing at dawn in the east. After dark the cosmos glitters gaudily, the planet Venus sharp and bright as a diamond in the west. A full moon spotlights the peak and throws a crisper shadow than in any city, where the light is diffuse and multi-angled, not one source in the sky. On moonless nights with thick clouds I can't see my hand three feet from my face. I step out the door of the cabin for a midnight leak, and mule deer drum down the hill, fading as they go, the drummers invisible in the squid-ink dark.

Being here alone I may not be my best self, in the social sense of the phrase, but I am perhaps my truest self: lazy, goofy, happiest when taking a nap or staring at the shapes of mountains. My friends back in town often want from me a report on the nature of solitude, on what it does to the mind to spend so much time alone.

In these early days, I'd have little to tell them. A few quiet words from Gary Snyder's "Lookout Journal" guide my days—

fewer the artifacts, less the words
slowly the life of it
a knack for nonattachment

My own insights are fragmentary, fleeting. I write something in my notebook and forget it an hour later. I do not so much seek anything as allow the world to come to me, allow the days to unfold as they will, the dramas of weather and wild creatures. I am most at peace not when I am thinking but when I am observing. There is so much to see, a pleasing diversity of landscapes, all of them always changing in new weather, new light, and all of them still and forever strange to a boy from the northern plains. I produce nothing but words; I consume nothing but food, a little propane, a little firewood. By being virtually useless in the calculations of the culture at large I become useful, at last, to myself.

SEEN FROM ABOVE ON A TOPOGRAPHIC map, my peak sits near the tip of an arrowhead-shaped convergence of mountain ranges. The arrowhead, encompassing a couple million acres of the wildest country in New Mexico, aims along a line parallel to the Rio Grande and points toward the Mexican city of Chihuahua. The Wahoo Range, the Black Range, the Mimbres Range, the Cookes Range, the Pinos Altos Range, and the Mogollon Range make up the angled edges of the arrowhead; its interior is filled out by the Diablos, the Jerkies, the Elks, the Lueras.

Off to my north it's all heavy forest and deep canyons dropping away from the rim of the Black Range, the heart of the Aldo Leopold Wilderness: sharp ridges, pink bluffs, dry arroyos, the occasional perennial stream fed by a spring on the lip of a cool *rincon*. Another lookout brackets the north half of the range, at Monument Mountain, and beyond him I can see both the Wahoos and a sliver of the grassy Plains of San Agustin. Far off to the northeast rise the great volcanic hulks of the San Mateo Mountains, former stronghold of Victorio, and just over their flank I can make out the Magdalenas.

To my south, the pine and fir forest of the high mountains eventually gives way to piñon-juniper foothills and finally to desert, with stark sky islands in the distance, the Floridas and the Tres Hermanas, hogback upthrusts of stubbled rock. To my immediate west are a series of ridges thickly covered in timber and brush, ridges that descend toward the valley of the Rio Mimbres. Along their tops can be seen the scars of old burns, grown back now in aspen, oak, and locust—none of which have yet leafed out, leaving the country in that direction looking rather drab, a charred snag looming up here and there like an iron spire. Beyond the Mimbres another string of mountains rises, the Pinos Altos Range. A lookout sits at its southeast tip, on Cherry Mountain, twenty-five miles west of me. He watches over my west-facing flanks, I cover his east-facing flanks, and we share a view of the valleys and mesas between us.

To the northwest is the roughest country of all, the headwaters of the West Fork of the Gila River. Out that way the mountain ranges crash upon each other like waves, each one higher than the next: the Diablos, the Jerkies, the Mogollons. A lookout occupies Snow Peak, the highest and most remote such post in all the Gila, just shy of 11,000 feet elevation and twelve and a half miles from the trailhead: the crème de la crème of wilderness lookouts. Its presence out there on the horizon evokes in me a perverse and loathsome envy, probably due to the fact that in all respects it is superlative to my own peak—bigger, lonelier, farther from the sun-seared asphalt, more intimate with the heavens. The two lookouts there, Sara and Razik, have a combined four and a half decades of experience on the mountain, and their best stories are better than mine—for instance the summer they were airlifted out by helicopter ahead of a fire that burned within twenty feet of their cabin. Slurry drops were required to save it, and they spent the latter portion of that season scrubbing out their cistern, the water in which had turned the color of pink lemonade from the slurry splashed over the cabin's roof.

The forest boundary is only five miles away to my east. Most of what I can see in that direction isn't forest at all, though occasionally it burns too—grass fires—and the private property owners and the Bureau of Land Management are as eager as the Forest Service to know when their turf catches fire. The Ladder Ranch makes up the greater part of the foreground to the east, a 150,000-acre swath of high mesas, broad canyons, and semi-arid grasslands owned by Ted Turner, one of the massive Western properties that make him possessor of more private land than any other American. East of the Ladder Ranch spreads the valley of the Rio Grande, running south at the base of the Fra Cristobals and the Caballo Range. Beyond the Caballos, farther east, is a brutal and mostly waterless stretch of desert known for centuries as the Jornada del Muerto, or Journey of the Dead Man. This ninety-mile length of the old Camino Real—the "royal highway" between Mexico City and Santa Fe—offered the traveler a relatively flat shortcut around the tortuous side canyons and up-and-down terrain of the Rio Grande drainage between present-day Las Cruces and Socorro. The trade-offs were oppressive heat in summer, little reliable water for humans and stock, and the ever present danger of Apache attack. The Jornada del Muerto took three days to cross if you didn't stop to rest at night, and, lacking water, most who took it did not.

Beyond the Jornada del Muerto I can see the San Andres Range, the mountains responsible for the white-gypsum sands of the Tularosa Basin, one of the natural wonders of North America. Sculpted by wind and water into sensuously curving dunes, some of them fifty feet high, the gypsum crystals on the east side of the San Andres spread across 275 square miles of nearly pure white desert, open to public visitation at the White Sands National Monument. The White Sands Missile Range sits just north of the monument. Off-limits to you and me, the missile range serves as a proving ground for some of the U.S. military's most advanced killing tools. It also contains that scarred piece of earth known as the Trinity Site, where man first unleashed on the world the splendors of the Bomb. Twice a year Trinity opens to visitors, who can inspect the desert for a new kind of rock, trinitite, which formed when the blast melted quartz and feldspar into green-colored glass.

In summary, then, I have a view of spruce-fir high country, ponderosa parkland, piñon-juniper hills, several river valleys, arid grassland mesas, dry arroyos, high-desert scrublands, and the occasional fire scar grown back in aspen and locust. Amid it all are contained the birthplace of the wilderness idea and the birthplace of the nuclear age—a landscape overlaid by history with equal parts hope and dread, and plenty enough irony to keep the mind at play through the long afternoons of no smoke.

As a lookout in high country, I like to tell people I get paid to look at trees. But the longer I work this mountain, the more I find myself thinking of grass.

For millennia, fire and grass worked symbiotically here.

Grass—bromes, gramas, muhlies, fescues, all the various bunch grasses native to the region—burned quickly and fertilized the soil, whence came fresh grass. Both lightning-caused fires and human-lit burns moved quickly through the forest understory. Trees lived in mature stands where most were hundreds of years old. Ponderosa covered much of the forest in open, savanna-like parkland, with trees widely scattered, surrounded by a sea of grass. Dendrochronology research—the study of tree rings for evidence of climate fluctuation and fire scars—shows that for centuries much of the forest burned once, twice, even several times a decade, until the late 1800s.

Then the sheep and the cow arrived in numbers previously unknown. With Geronimo's surrender in 1886 the last of the Apache holdouts were finally subdued, and livestock barons pushed their herds into the most remote corners of the Southwest, searching for good grass on what was left of the open range. Hundreds of thousands of head of livestock swarmed over southern New Mexico, up to the highest reaches of the Gila country. The consequences were shocking and immediate. In 1895, a great flood swept out of the Pinos Altos mountains and tore through what was then the heart of Silver City. Subsequent floods wreaked even more havoc. Homes, businesses—even famously a grand piano owned by the justice of the peace—were washed away in the periodic surges. The town's commercial strip had to be moved one block west after the original Main Street became a fifty-foot crevasse, the floods having scoured the earth to bedrock. That scar, colloquially called the Big Ditch, can still be seen today.

With the coming of the Forest Service at the beginning of the twentieth century, two attitudes prevailed: deference to ranchers and disdain for fire. In 1826 the explorer James Ohio Pattie had reported being "fatigued by the difficulty of getting through the high grass, which covered the heavily timbered bottom" of the Gila River drainage. By 1909, forest ranger Henry Woodrow noted that in the heart of the Gila, at what today is the point in New Mexico farthest from a road, "There were cattle all through the mountains and grass was hard to find." He had to ride his horse a dozen miles to find something on which it could graze. Elsewhere on the forest sheep were driven from grassy mesas to water sources in the valleys, carving trails where every living thing was trampled; the trailways acted as barriers to the spread of fire. By the middle of the twentieth century, much that was wrong with the Gila could be summed up in six words: too many livestock, too few fires. The forest had become a scenic pasturing ground for private ranchers, as indeed too much of it remains.

On this day, though—tax day, April 15—I need only look out my windows to see how far we've come in embracing fire's utility. A line of smoke rises just this side of North Star Mesa, on land once grazed by the GOS Cattle Company, one of the biggest ranching outfits in the history of the Gila. Instead of cows converting grass to protein, fire converts it to ash. Instead of an incident commander marshaling resources to squelch it, this fire—a "prescribed fire"—is led by a burn boss. The burn boss directs the movements of men and women walking along a pre-mapped perimeter with drip torches, canisters that pour a mixture of diesel and gasoline out of a nozzle and past an igniter, allowing their users to dribble flames on the ground. The goal is for the fire to catch and begin to move east, burning several thousand acres over the course of a few days, mimicking an ecological function removed from the land for most of the twentieth century. It takes a good deal of study and some imagination to see in the mind the mesa as it was once: a lake of grass rippling in the springtime wind. Partially colonized now by juniper and various forms of woody scrub, it little resembles the state in which it was first seen by whites and Hispanics. Fire, so the hope goes, will begin to change that—fire in its ancient marriage with grass, the two of them conspired against brush.

Fittingly, it was Aldo Leopold who first began to tease out the connections among grass, brush, and fire. In 1919, ten years after arriving in the Southwest, he was appointed assistant district forester on 20 million acres in Arizona and New Mexico—a job that involved oversight of personnel, finances, fire prevention, and roads and trails. He inspected the forests up close, taking long horseback rides on all the reserves in both states. He devised a record-keeping system for inspection tours and wrote detailed memoranda of his findings. What he saw troubled him.

Changes could be read everywhere on the land. On the Blue River of the Apache National Forest in Arizona, for instance, widespread erosion had torn out the lush river bottom where settlers once farmed. Trout streams had become waterless ribbons of cobble bar and driftwood. Marshes had been drained by the cutting of gullies. Two questions presented themselves to Leopold. What had caused the changes, and what could be done to mitigate them?

The prevailing ideology of the Forest Service prized mature and loggable timber above all, measured in board feet and dollars. Grazing went hand-in-glove with timber production; it kept the grass down, limiting the spread of fire. Fire took potential money in the form of timber fees and sent it to the sky in a puff of smoke. With fire removed from the scene, new trees took hold, trees that could one day be cut. Cows and sheep, in this formulation, helped make the timber ripe for the harvest. Two sources of revenue, mutually reinforcing: grazing fees and timber fees. A government bureaucracy's dream.

Something had gone wrong, though. On his first reconnaissance mission in the Apache, in 1909, Leopold had found 300 people homesteading on the river bottom of the Blue. On his return twelve years later, ninety people remained. The riparian ribbon of life along the river was desiccated. Willows had been torn out by violent floods. Good soil, each inch of which had taken as much as a thousand years to build, had washed off the hillsides. Gullies scarred the slopes. Lacking competition from surface fires flashing through the grass and snuffing saplings, trees and chaparral were spreading on the benchlands and foothills. More timber, perhaps—but at what cost? Leopold spent weeks at a time in the field. He saw firsthand the condition of the land, and everywhere he went he looked hard for answers. Grazing appeared the obvious culprit.

Through a series of extraordinary reports and articles in the early 1920s, we can follow the evolution of Leopold's thinking as he put forth theories to explain what he saw. In 1920, he wrote dismissively of the notion that fire could have any positive impact. "Piute Forestry," he sneeringly called it, a form of primitivism and an insult to his scientific cast of mind. In the course of a few pages in *Southwestern Magazine,* he got everything we've subsequently learned about fire ecology dead wrong. Four years later he was a little less sure of himself. He had studied tree rings and burned-over juniper stumps—becoming a pioneer in the use of dendrochronology to determine fire history—and was puzzling over whether grasslands were the "climax" vegetation, aided by fire, or just a phase through which the land passed on its way to a different climax of deciduous scrub and dense brush: oak, manzanita, mountain mahogany. He got this wrong too, at least for a while. But of the changes and their cause he had no doubt, as he wrote in the *Journal of Forestry:*

43

Until very recently we have administered the . . . Forests on the assumption that while overgrazing was bad for erosion, fire was worse. . . . In making this assumption we have accepted the traditional theory as to the place of fire and forests in erosion, and rejected the plain story written on the face of Nature.

In an article that same year for *Sunset* magazine, he went so far as to note that in an arid climate such as the Southwest, any grazing at all, even of the most conservative kind, would likely produce erosion. For a high-ranking member of the Forest Service, this amounted to apostasy. To this day, stubborn advocates of public-lands ranching refuse to hear of it.

Needless to say, the answer is not as simple as removing all the cows and torching the mesas they once grazed. One fire won't be enough to adequately thin the spreading brush and juniper; fire will need to be a recurring presence, and it remains to be seen if both the civilian public and the Forest Service have the stomach for it over the long run. Prescribed fires don't always obey the boundary of their prescription, the most famous local example being the Cerro Grande Fire of 2000. Begun as a prescribed fire in the Bandelier National Monument of northern New Mexico, it escaped control and burned for a month before it was contained. The smoke plume blew so far east it could be seen over the Oklahoma panhandle, 200 miles away. Forty-seven thousand acres burned, causing a billion dollars' total damage, including $340 million worth to the Los Alamos National Laboratories. The habitat of endangered spotted owls was incinerated. The town of Los Alamos had to be evacuated. More than 350 families lost their homes. Officials quibbled for years over what had gone wrong, and why. Aside from the quantifiable destruction of property, the biggest loss may have been public confidence in the very notion of prescribed fire as a legitimate tool for ecosystem restoration—at least in the intermix zone of wildlands and homes. It's hard to win a friend for fire when you burn down his house with it.

The Gila, though, provides a canvas on which to paint with fire that is superior to any other place in the Southwest. Indeed, if it can't be done here, it probably can't be done anywhere in the Lower Forty-eight. Only a few tiny towns are sprinkled throughout the forest. The vast majority is uninhabited: less than one-half of one percent of it qualifies as wildland-urban interface. The biggest town around, Silver City, with a population of 10,000, lies at the Gila's southern edge, and with prevailing winds out of the southwest, the smoke impact on humans is close to negligible. None of this means precautions aren't taken, down to the tiniest detail. As the writer William de Buys has said: "Consider the alignment of stars a burn boss on public land must achieve: archaeological clearance, interagency consultation on threatened and endangered species, environmental analysis, survival of appeals, clean air permit, crew and equipment availability, weather window. A thousand-acre burn can take a year or more of preparation. . . . Even then its chances of occurring in optimal conditions are slim."

On this mid-April day in the Mimbres Valley of the Gila National Forest, the stars have aligned. Winds are light. Fuel moisture is advantageous: neither too dry, which would make a burn dangerous, nor too wet, which would make a burn pointless. The fire goes off without a hitch, several hundred acres a day, and from where I sit it even looks rather beautiful, a soft white smoke rising off the tablelands, blown east over the Black Range crest before dispersing on the desert.

. . .

IN THE THIRD WEEK OF APRIL, the snowpack having receded, the packers Les and Kameron arrive with the boxes I stowed in the shack at the saddle. They tie their animals to various trees in the meadow, and together we empty the mules' panniers of my supplies. When I shake Les's hand and ask how he's doing, he smiles through his neatly trimmed beard and says, "Living the dream, man. Living the dream."

Hard-handed and bow-legged, these men practice one of the great and dying arts of a world without roads, and over the course of my season their handiwork today will lessen the punishment my body takes in getting to the lookout. With a base of dry and canned food in place, I will resupply myself every two weeks with fresh fruit and vegetables, bread, chocolate, cheese, and tortillas. Their effort today is rather simple, just two men on horseback, each of them leading a single mule on a rope, with panniers holding my boxes balanced at each mule's side. I have seen them supply a big crew in rough country, however, where a single man may lead a string of as many as ten animals, and it is a sight of calm and purposeful synchronicity sweet to behold, not least because the mules seem to enjoy it.

We stack my boxes on the porch and share lunch in the warmth of the cabin. Les has enough food for three men, a feast put together last night by his wife: sandwiches, yogurt, cookies, homemade trail mix. He graciously offers me his leftovers. I consider most of what he brought a luxury here. I do not turn it down.

"You ever get lonely up here?" Kameron, the younger of the two, asks as he looks around the cabin.

"Nope. Not really."

"You ever get, you know, sad or anything like that?"

This requires a thoughtful response. Sad is too small and mean a word for the feelings this place evokes in me, yet I'd be lying to say I live here in a state of perpetual ecstasy, like the blissed-out bodhisattva Kerouac dreamed of becoming. On the other hand I'm not about to tell two guys in leather chaps and cowboy hats about my very real and near-mystical hours of longing and nostalgia, alone in my little glass box, brooding over and exulting in my own mortality amid mountains silently magisterial in the late-day sun. Nor the hours of sitting and staring into the inscrutable heart of the desert, not thinking anything, not feeling anything—neither happy nor sad nor any quantifiable mixture thereof. Merely alive with a hungry retina and a taste for dry mountain country and a jones for the sight of that first twist of smoke.

"The only time my heart sinks is when hikers show up. But it doesn't happen often and they never stay long."

They laugh and look at each other, wondering, perhaps, if I'm trying to tell them something. It's no wonder our Forest Service brethren think of us lookouts as the freaks on the peaks. We have, in the words of our forebear Edward Abbey, "an indolent, melancholy nature." Our walk home is always uphill. We live alone on the roof of the world, clinging to the rock like condors, fiercely territorial. We ply our trade inside a steel-and-glass room immaculately designed to attract lightning. Our purpose and our pleasure is to watch: study the horizon, ride out the storms, an eagle eye peeled for evidence of flame.

Kameron leans back and yawns, rubs his clean-shaven jaw. The clouds have begun to spit sleet, compact little crystals flying horizontally on the wind, even as the sun shines overhead. Though it's just a little squall, unlikely to last long, it serves as their excuse to get moving down the trail. I stand in the meadow and watch them descend through the trees. Les turns, offers a wave of his hand, and then they are gone. If I'm lucky, they'll be the last of my government colleagues to set foot on this mountain all year. From here on out the peak will be mine. The thought makes me smile, as does the knowledge that hidden in my boxes of books are three spare bottles of bourbon.

ON THE LAST MONDAY of the month I hike in for the start of another tour, smelling along the way the sweet dry smell of pine needles in sunlight, the dog fifty yards ahead of me, both of us loafing along, finding our rhythm—no reason to rush on such a fine spring day, a beautiful morning in the Black Range for April, unusually calm. We're a mile up the trail from the saddle, just crossing a short ridge connecting two hills, when I see a globule of smoke curl out of a canyon bottom, a mile northeast and a thousand feet below.

Fire.

Several thoughts occur to me in rapid succession. First, there is no relief lookout as yet on the mountain. On my days off— and I've been working overtime thus far, twelve or thirteen days instead of ten, with only a day or two off at a time—the lookout remains unstaffed, my boss having hired as my relief a student who won't be free until the last week of May. The absence of an upcanyon haze of drift smoke indicates the fire has just popped up. The color and the shape of the smoke tell me still other things, so by the time I drop my pack and key up my radio—Alice looking at me curiously, her head cocked to one side—I'm pretty certain I'm the only person on earth who can see it. I also have a good idea of where it is, how it started, how big it is, and what sort of fuels it's burning in, a bit of guesswork I can indulge after eight seasons in this line of work.

Although my knowledge of the ways of fire has grown, my initial reaction to the sight of a smoke remains unchanged over the years. It's always a major kick. I take a couple of minutes to regulate my breathing and douse my sizzling nerve ends in the cold water of government professionalism before I press the transmit button on the radio. Otherwise I'd sound as if I were the thing on fire.

Out of habit I call dispatch under the name of my peak instead of my last name, which causes them confusion given my location. More than once they request my azimuth reading on the fire, before I finally get them to understand I am calling from the trail and not from the top. I tell them the fire is about a mile north-northeast of Wright's Saddle, that it's on the east side of the Black Range, in a canyon bottom, about a tenth of an acre or less, burning most likely on the ground, in grass, a small column of light gray, almost white smoke. With four and a half miles ahead of me, I tell them to give me an hour and a half to get to the top, at which point I'll be happy to give them my azimuth from the tower. Then I switch channels and call my boss to make sure he has a copy on all I've just said, and by talking it through—him back in the office with maps, me standing on a boulder looking into the canyon—we put a pretty firm location on the fire: a little ways beyond the end of a road heading west out of Embree, near where trail turns up Thief Gulch. Thus, the Thief Fire.

If you threw a hundred darts at a map of the Black Range, you'd be hard pressed to hit a spot more conducive to an unwanted fire. Brushy fuels cover steep slopes at the confluence of three drainages, offering the fire ample combustibles and plenty of avenues to move. Fire has been ruthlessly suppressed in the surrounding country for a hundred years, so fuel loads are unusually high. The rough terrain makes the area too dangerous to work with a helicopter, removing one tool from the suppression arsenal. The proximity of Embree, less than two miles east, adds private homes to the list of concerns. Were the fire to spread upslope to the west, it would begin to find thick pockets of dead fir perfectly poised to burn. Beginning in 2003, engraver beetles began to attack white fir stands on the southern end of the Black Range, resulting in an almost total mortality for that tree species over tens of thousands of acres. During my tenure I've seen the needles of the firs to my south turn from healthy green to bright red, then become ashen gray and fall to the ground, leaving dead standing trees to weather and rot.

Such beetles are endemic in Western coniferous forests. The mountain pine beetle, for instance, attacks lodgepole pines and is responsible for decimating nearly 4 million acres of Western pine forests in 2007 alone. Add fir beetles and the tally comes closer to 7 million acres. Bark beetles typically feed on trees already weakened by other factors. Occasionally, when forests become stressed by drought, overcrowding, or widespread root disease, beetle populations erupt into epidemics. This has happened on an irregular schedule for thousands of years; it happened here over the past half decade. A forest overgrown from an absence of fire became further stressed by a period of drought: too many trees competing for too little moisture. The beetles found thick, susceptible stands of white fir and bored through their bark. They laid eggs in little galleries carved in the space between the cambium and the sapwood. The eggs hatched and became larvae, which fed on the cambium, the living tissue that cycles nutrients from the roots to the rest of the tree. After a year, the larvae emerged as full-grown beetles and flew away to attack new trees. Only the eventual decimation of the host tree species can halt an epidemic—either that or a late hard freeze that kills the larvae. No such freeze has occurred here in years, so the beetles had their way until they killed most of the white fir trees over an area bigger than the island of Manhattan.

About 60 percent of the more than half-million-acre Black Range District has seen fire over the past fifteen years—all of that to my north. In 2003, for instance, the Boiler Fire burned 58,000 acres. In 2006 the Bull Fire burned 80,000 acres. Such figures include the total perimeter of a burn; inside the perimeter many pockets of green remain untouched by flame, and even in the charred places big-boled trees live on. These fires—the kind that burn slowly, a few hundred to a couple thousand acres a day over a period of weeks or months—create a mosaic of habitats that in their overall health and diversity help prevent bark beetle attacks. The one area of the Black Range still starved of fire is the country running from my lookout south to the forest boundary, precisely the area with the worst beetle infestation on the entire range.

It's not difficult to grasp why fire was kept at bay here for most of the twentieth century. There is, first of all, the presence of homes and cabins along the forest edge. A scenic highway passes by, and the tourists prefer green trees to black. The mountains here hold the watershed of two little towns beyond the forest, each of them perched astride the creek banks, and a fire would strip ground cover and result in rapid muddy runoff right down their main streets. The public angst likely to arise from the sight of flames and mud was judged not worth the risk. In my time I've seen crews suppress at least half a dozen fires that could have grown huge to my south if left to burn.

The consensus now, from the district ranger on down through the firefighters and even to us lowly lookouts, is that this corner of the forest is going to burn, has to burn, one of these years. It's only a matter of how and when. Maybe we'll get lucky with a late-season smoke that hangs on through cool weather and even light rains, burning calmly for weeks, chewing up the deadfall and duff on the ground, torching through occasional dog-hair thickets of pine and fir, and leaving the mature forest partially intact. On the other hand, when the big one comes in this part of the world, the stage is more likely to be set by high winds, hot summer weather, and low humidity—followed by ignition from dry lightning or the careless human hand. A fire that can't be caught by anything but rain, no matter the manpower thrown at it. We see such fires with increasing frequency in the Southwest: the Rodeo-Chediski and Aspen fires in Arizona (467,000 and 84,750 acres, respectively), the Hayman Fire in Colorado (137,000 acres), the Bear Fire here on the Gila (51,000 acres), fires that blew through 10,000 acres a day or more and torched every last living thing they touched, leaving a charscape in their midst.

However it happens, I want to be here for it, partly out of a selfish wish for spectacle, partly because I've stared at the land for thousands of hours and feel I could be useful to the folks on the ground.

WITH FIREFIGHTERS COMING FROM BETWEEN forty and a hundred and twenty miles away—no smokejumpers available in the Southwest just yet—I arrive on the peak about the same time the first crew of five reaches the Thief Fire. An airplane observer is up and circling, talking to the crew on the air-to-ground frequency. From the tower, I'm just able to see the top of the smoke rising above a ridge. It is my hunch, and it's only a hunch, that if I'd been on the peak instead of hiking in, I'd have been a half hour later in seeing the smoke. Pure dumb luck to have been where I was, when I was.

The crew's first task is to locate the fire's origin and determine a cause, whether lightning or human. Much hinges on what they find. If it's lightning-caused, my boss has several options. He can order suppression. He can put the fire in confinement status and wait to see how it behaves, planning to corral it several days out. Or he can call it a "fire-use" fire and let it burn for weeks or months on end. If, on the other hand, the fire is found to be human-caused, his only option will be total suppression.

For most of the twentieth century there was no such palette of options. The Forest Service judged each and every fire a potential travesty and responded accordingly. This policy can be traced back to the agency's earliest days. In 1905, when President Theodore Roosevelt brought the nation's forest reserves under the control of the Department of Agriculture, he appointed his friend and fellow conservationist Gifford Pinchot as the first chief of the U.S. Forest Service. Pinchot, like Roosevelt a political Progressive, believed the Western forests ought to be protected from the baleful forces that had laid waste to the forests of the East and Upper Midwest—clearcuts and fires, often in that order. The goal was not to preserve timbered lands as semi-wild getaways for that portion of the American public that liked dramatic scenery. That fell to the national parks. Rather, it was to impose a Progressive-era ideal of efficiency and rational use on the raw materials of empire.

Throughout the second half of the nineteenth century, timber barons, railroad trusts, and mining concerns had mowed down vast reaches of mature timber to feed the nation's hunger for wood products, whether for housing, railroad ties, mine shaft supports, or fuel. Forestalling these interests, powerful as they were, was the easy part; moving the forests to the public domain did much to inhibit their destruction. But one force proved a trickier enemy: fire. It had been a presence on the landscape far longer than big trusts and rapacious conglomerates. Free-burning fire, kindled by lightning, had been the norm in Western forests for untold millennia. Its intensity fluctuated depending on variations in fuels and climate, flaring in drier epochs, fizzling in wetter ones.

But as the fire historian Stephen Pyne points out, wildfire has been a presence on Earth since the Devonian period, 400 million years ago, when sufficient oxygen for combustion flooded the planet's atmosphere.

Once humans mastered fire, they wielded it liberally for their own purposes. They used it while hunting, to herd, confuse, and entrap their prey. They used it for farming, to clear the land of cover and fertilize the soil for cultivation. They used it in tandem with domestic livestock, to jump-start regrowth on previously grazed grass. They even used it for gathering purposes, aware that light burning made way for acorns, wild raspberries, and other food sources. Sometimes their fires escaped and burned over areas larger than intended. The combination of lightning-kindled wildfire and human burning created a crazy quilt of flame-licked lands.

So the landscapes gathered under the newly born American forest reserves had long been the site of a complex fire regime. With the coming of European civilization, fire's presence only intensified. Yeomen farmers practiced slash-and-burn agriculture, dumping brush and grass back into the soil as ash and breaking up thick root systems in preparation for seeding with grain crops. Railroads threw sparks from their tracks and embers from their coal-fired engines, showering whole corridors of the West with fire. Loggers torched their leftover slash. Drifters, settlers, and prospectors lit fires by design and by accident. Campfires escaped. Mining towns burned into the surrounding hills.

To early Progressive foresters, all this fire on the land showed a promiscuous disregard for the nation's timber resources. Bernhard Fernow, Pinchot's predecessor in the Division of Forestry, considered this flowering of fire a result of "bad habits and loose morals." Pinchot equated fire suppression with the earlier crusade for abolition: "The question of forest fires, like the question of slavery, may be shelved for a time, at enormous cost in the end, but sooner or later it must be faced." On an inspection of Arizona forests in 1900, Pinchot rode horseback along the Mogollon Rim, looking over North America's largest continuous stand of ponderosa pine. "We looked down and across the forest to the plain," he later wrote in his memoirs, "and as we looked there rose a line of smokes. An Apache was getting ready to hunt deer. And he was setting the woods on fire because a hunter has a better chance under cover of smoke. It was primeval but not according to the rules." It didn't occur to him that what was primeval and what was the rule might be one and the same, and for good reason.

The great fire debate commenced sooner than even Pinchot anticipated. In the summer of 1910, five years after Roosevelt had created the Forest Service, and a mere three years after he and Pinchot had added an additional 16 million acres to the national forests in a sweeping exercise of executive power, fires broke out all over the country, from New York to Oregon. The worst fires flared in the northern Rockies. In the national forests of Montana and Idaho, nearly 3 million acres burned—and that doesn't count fires in the surrounding national parks and Indian reservations, much less on private holdings or state-owned lands. Seventy-nine firefighters died fighting the biggest blazes. Whole towns collapsed in flame. Some estimates put the total burned acreage for the year in the tens of millions nationwide.

Three main arguments dominated the debate that followed. Some claimed the answer lay in light burning—the Indian way, it was sometimes called. In an article published in *Sunset* magazine the month the great fires blew up, three civil engineers made the case for "how fire must be fought with fire." They pointed out that California's pine forests had been burned by natives for centuries before white men arrived, and what white men found were healthy forests with open, grassy understory and accessible timber. By continuing such practices, foresters could make fire their servant rather than their master—lighting fires when conditions were favorable, in the cooler, damper weather of autumn, rather than fighting fires in the searing heat and squirrelly winds of summer. Preemptive use of fire would protect against the kind of blowups just then charring the northern Rockies.

For those who opposed federal control of the forests, the catastrophic fires reinforced their belief that settlement was the only appropriate response. Development would reduce fire danger by threading the forests with roads, chopping the vast stretches of continuous fuel into more manageable chunks; by opening the timber to cutting, which would further starve fire of fuel; and by populating the mountains with settlers, who could organize themselves to fight fire when it occurred. To the pro-development forces—and they included many powerful Western congressmen—the enemy to be feared wasn't fire, it was conservation. Senator Weldon Heyburn of Idaho claimed the big fires were God's revenge for not allowing industry to have its way on the land. He pushed a plan to remove the burned areas from national forests and hand them over to private owners. The *Idaho Press* went so far as to suggest clear-cutting the northern Rockies as a defensive measure: "It would really be better to cut down all the trees than to incur the imminent risk of such vast destruction and mortality as has been accompanied by these fires. For it is better to devastate forests than to devastate settlements."

Pinchot and the Forest Service pushed back. They viewed such arguments as a means of undermining public confidence in the agency, which still faced opposition in Congress and an annual battle for appropriations. To usher in settlement or approve wholesale clear-cutting would defeat the very purpose of the national forests. To admit the efficacy of light burning would be to capitulate to primitivism, confusing the public. The Forest Service existed to preserve the public good, and fire destroyed the public good. What was required, in Pinchot's reckoning, was a military-style assault on fire. Seventy-nine men had died in Idaho and Montana not for a misguided cause; they'd died because obstructionists in Congress hadn't provided them with adequate resources. With a system of trails, telephone lines, and lookout towers in place in the national forests—plus the promise of emergency appropriations to cover firefighting costs—early detection and rapid response would give firefighters a decisive advantage. Fire could be banished, relegated to the history books.

This argument won the day, prevailing for sixty years. It had the virtue of simplicity: all fire in all forests was evil and must be stopped. It reassured the public that the men who'd perished in the big blowup had not died in vain; in fact they were the first martyrs in a noble crusade. With the trauma of 1910 seared in its DNA, the fledgling Forest Service had a defining purpose around which to rally. Dissenting voices would be marginalized for decades to come as the agency deployed its shock troops throughout the West, to wage its ruthless war on fire. The fact that roads, trails, telephones, and towers constituted their own form of development did not trouble the sleep of Pinchot. He had always said that "the first great fact about conservation is that it stands for development." To him, there were two kinds of development: the kind that destroyed the forests for short-term private gain and the kind that preserved them for the long run. Fighting fire would preserve them for the long run.

Decades would pass before the folly of this approach was finally acknowledged. Trails, phones, and towers would be supplemented with tankers, helicopters, and bulldozers, and the cost of fighting fires eventually would run beyond a billion dollars a year.

AT A LITTLE PAST NOON, the crew returns its verdict on the source of the Thief Fire: an abandoned campfire, no surprise given the absence of storms and the fire's location on a trail in a creek bottom, not far from a dirt road. Perhaps a couple of weekend turkey hunters broke camp hurriedly, confident a stone ring would hold their fire's embers in check. No such luck, not this time. I have camped in heavy Black Range downpours, managed to get a campfire going, and found hot coals from which to rebuild it in the morning despite overnight showers. In the weather we've had lately—warm and windy, following on a dry winter—a pit of ash over coals can smolder for a couple of days, heating the surrounding earth until the pine duff around the stone circle ignites.

At 12:40 Tony, the incident commander (IC), calls his size-up in to dispatch. The fire is ten to twelve acres now, burning in grass and piñon-juniper, smoldering and making isolated runs, with flame lengths of two to three feet. Winds are light and variable, gusting to seven. Values at risk are private property holdings: the community of Embree, as well as some old mining shacks to its west. Potential for spread is high. Tony requests a heavy tanker for slurry drops and a second hot-shot crew in addition to the one already on order. After signing off with dispatch, Tony dials up his crew on one of the tactical channels I've reset my radio to scan, in order to eavesdrop. "Jimbo's gonna mark off the fire's origin with some flagging tape. Make sure you've got plenty of water. I've got reinforcements coming. We're gonna need everybody to be heavy hitters on this one."

By midafternoon seventy pairs of boots are on the ground—three hot-shot crews and two engine crews—working furiously with chain saws, Pulaskis, shovels, and rakes to cut and dig a containment line down to mineral soil along the fire's flanks. The goal is to halt the fire's spread by starving it of fuel, keeping it inside the line. The observer plane circles overhead, giving the IC updates on the fire's behavior. The observer-plane pilot requests a TFR—total flight restriction—over the area, to 3,000 feet above ground level, in order to keep civilian and military aircraft clear of aerial operations over the fire. Guided on its drops by another aircraft called a lead plane, the tanker makes runs with slurry over the ridge tops, hoping to hold the head of the fire in check long enough for the ground crews to bring their lines along both flanks and tie them into the slurry. Every now and then a stand of thick timber torches, and a dark gray cloud of smoke rises into the sky.

I turn away from the fire momentarily, undertake my periodic survey of the country in the other direction, to the north and west. I'm scanning the Pinos Altos Range through the binoculars when something big and black takes up half my view. It's on top of me almost before the sound of it arrives: an air force fighter jet. I don't even have time to offer my customary one-finger salute. I call the dispatcher, as is the protocol in such situations.

"Silver City Dispatch, Apache Peak, I've got a visual on a military jet headed due east over Apache Peak."

"Apache Peak, dispatch, we copy. Thanks for the info."

I switch to air-to-ground, call the observer plane.

"Heads up, John, my tower just got buzzed by a military jet."

"Yeah, we copied direct. . . . In fact, we've got him in sight. Stand by one."

While I'm standing by, the observer plane is forced to take evasive action. The pilot dives and turns 180 degrees. The wing man identifies the incoming aircraft as a delta-wing single-exhaust Tornado jet: hot-dog flyboys, out on a training run. They love to fly the Gila. It's wild, it's rugged, hardly anyone complains of the noise because hardly anyone lives out here. I see them a couple dozen times a summer. Somehow, though, someone didn't get the message on the TFR, and the results are nearly fatal—certainly way too frightening to mention in any detail on the radio, though no doubt the pilot and the spotter uttered choice words in that cockpit.

"Apache Peak, Thief Air Attack."

"Thief Air Attack, Apache Peak."

"Thanks for the heads-up, buddy. That got real Western there for a second."

"Copy that. I'd say he came within two hundred feet of my tower."

"No excuse for that. Or for blowing the TFR. We'll make sure they get an earful over at Holloman [Air Force Base] when we get back in tonight."

Given how the culture of wildland firefighting, almost from the beginning, has borrowed its tools and tactics from the American military, it's no small irony that on this day a rogue air force jet provides the nearest brush with disaster.

AROUND FOUR O'CLOCK THE THIEF Fire crosses the main creek bottom and begins to burn up the other side. The IC admits to the observer plane that despite their suppression efforts they may not catch this one. "Don't know what to recommend to you," the spotter replies. "You're doing all the right things, but it's kind of a dog's breakfast down there."

The IC calls me next. We know each other well, Tony and I. He's my boss's second-in-command, has been on the Gila five years. No need to mince words. We both know the score: If the fire continues to the top of the ridge on the north, it will eventually find continuous brush and heavy timber for miles to the west, including all the way to the top of my peak.

"Phil, have you got your safety zones planned out?" Tony asks.

"I've got options. The meadow is big and open and mostly grass. There's also a little boulder field where the deer have munched away all the vegetation. And of course there's always the cistern. I could drop down in there and tread water for a couple hours if it came to that."

"Well, we'd try to fly you out by chopper before then. But just think ahead a little for me. Safety's absolutely number one."

"Copy that," I say. "Good luck down there."

In addition to the options mentioned above, I have three others. I could simply stay in the tower, which rises far enough above the meadow to protect me, unless the trees to my north torch in flame lengths of fifty feet or more (always a possibility), in which case I'd roast in the tower like a chicken in an oven; I could light my own backfire, burning up enough of the meadow ahead of the main fire to rob it of fuel and allow me to lie down in the ashes; or I could run to the cabin for my fire shelter, a kind of miniature tent of aluminum and fiberglass that reflects radiant heat and traps breathable air, allowing a human to survive inside of it while a fire passes over. None of these options is terribly attractive, and each would be a last resort in case a helicopter couldn't come for me. But with three and a half miles between me and the fire, and prevailing winds working against it, the danger is far from imminent. I'll say this for my superiors: they mean it when they talk safety first. Last year, for instance, twenty-six firefighters died in the line of duty on wildland fires in the United States. It's an inherently dangerous line of work, and planning for maximum safety is always my bosses' top concern.

At 5:15 p.m. a second tanker arrives on scene, and for a couple of hours the two tankers trade off, working load and return, load and return out of Grant County Airport in staggered increments, dropping 3,000 gallons of retardant with each load. Designed to coat vegetation in moisture, the retardant contains 85 percent water, 10 percent sulfate-based ammonia fertilizer, and a small amount of iron oxide to color the mixture red, so the pilots can see their handiwork. All radio communication on the fire is punctuated by the whine and roar of chain saws in the background. When the observer plane returns to the airport for fuel, I take over as eyes from above on fire behavior. The crews continue to build line on the fire's flanks. It's my job to let them know if the color and volume of the smoke change. Every now and then it does, as fire finds brush and trees to torch; the crews report several slopovers, meaning fire has crossed their line and new line must be scraped to contain it.

As dusk begins to fall, the fire cools amid rising humidity, falling temperatures, lighter winds. This is the moment of truth. With one last desperate push, the separate crews tie their lines into the slurry barrier, and the fire's growth is halted at twenty-five acres. There will remain several days of mop-up and monitor duty, not to mention a long night of patrolling line and cooling off hot spots on the edges, but the battle, this time, has been won. The IC thanks the tanker pilots for their work and cuts them loose. The observer plane bids farewell, turns back toward the fire base. "You saved our bacon on this one," the spotter tells the tanker pilots. "That could have been twenty thousand acres if things didn't go our way. Let's do it again sometime, but maybe not real soon."

No injuries, no damage to property, no explaining to prying reporters why this one got away: all in all the best possible outcome, at least from a public relations standpoint. The residents of Embree will sleep easy tonight, and once the crews are back in camp and bedded down around midnight, so will I.

2

❈

MAY

❈

I was expected to sit still and watch mountains and long for company and something to do, like playing cribbage, I suppose. I was going to have to watch mountains for sure, that was my job, but I would not be without company. I already knew that mountains live and move.

—Norman Maclean, "USFS 1919"

*The pleasures of Frisbee golf ✳ a heli-
copter beer run ✳ cowering before the
wind's wild power ✳ an encounter with
the Electric Cowboy ✳ trout fishing
by moonlight ✳ Black Larry's Rules
for Black Range Travel ✳ a love story
✳ smokejumpers on the Cobre Fire ✳
the brief career of one relief lookout*

I LOVE MY OFFICE. TWENTY paces from the cabin, sixty-
five more up the steps of the tower, and just like that I'm on
the job. With my housekeeping complete, my supplies put
up, and a good stack of wood on the porch, I begin more or less
full-time service here, 9 a.m. to 6 p.m., an hour off for lunch—a
schedule not unlike that of any other jogger on the hamster wheel
of the eight-hour day.

Many visitors think I work both night and day, assuming fires
are easier to spot after sunset, with bright flames dancing in the
dark. It is true that if a fire gets going, it can sustain itself overnight for days or weeks. In high winds or drought conditions it

may even grow substantially at night. But wildfire thrives in warm air and low humidity—conditions found between the hours of 10 a.m. and 4 p.m., roughly speaking—and it usually shows itself not with open flame on the ground or in the crowns of trees, but in tendrils of smoke drifting up and curling, perhaps even puffing and then disappearing for a while, a kind of taunt or tease that leads a lookout to reach for the binoculars and keep coming back to a single swath of country.

Alice likes to laze in the shadow cast by the room where I sit, and from her angle of repose she acts as a lookout in another form. As I watch for fires, she keeps an ear to the ground for the appearance of hikers and barks to let me know when someone's coming. The curve of the north slope acts as a funnel on sound, and hikers approaching from below disturb the sonic peace just enough to jolt her from her slumber. She rarely barks in her other life as a town dog, but here she tends toward the territorial, as if out of a race memory of guarding camp in far-off Labrador. With her warning howls to alert me, I'm never caught napping on the job.

Over the years I've made some improvements to my spartan work space. With a straight length of pine limb and a square of plywood, I've fashioned a writing table wedged into one corner of the tower, just big enough to hold my typewriter. It allows me to write while standing up; in this way I can type and look out at the same time. Along the east wall of the tower, between the fire-finder cabinet and the east-facing windows, I've rebuilt a rudimentary cot, a human-size slab of plywood perched on legs cut from an old corral post. Made up with a sleeping pad and a Forest Service bag, it offers ample comfort on which to read and allows me to look out merely by sitting up.

Once I would have used my early-season freedom to compare my maps with the country out the windows, putting names to every creek, ridge, peak, or mesa on my turf, preparing myself to pinpoint smokes. But I've been here long enough to know the landscape pretty much cold, at least that portion of it for which I'm responsible, a jagged radius of about twenty-five miles in all directions. The game now, when I choose to play it, involves scanning the far horizon for the most dimly visible shapes on the clearest day, 150 or 180 miles away: ranges in Texas and Arizona and far northern Mexico, the Franklins and Pinaleños and Sierra Madre Occidental, faint thin smudges on the edge of my world. If the view's not great, I sometimes scan the big forest map, adding to a list I keep of the more colorful place-names on the Gila: Granny Mountain, Noonday Peak, Booger Canyon, Poverty Creek, Spud Patch Ridge, Jackass Park, Deadman Spring, Sixshooter Saddle.

But not today. Today is May Day, the international day of worker solidarity, and in honor of my hard-toiling brethren everywhere I choose to remain prone with a book. On the clock, to be sure—a paid holiday of sorts. As it should be.

Now and then I lift my head to watch the local bird life. Turkey vultures soar and swoop on the thermals, a fairly constant presence here. A red-tailed hawk whistles on the north slope. In the afternoon, three dozen violet greenback swallows circle and dip over the ridge to my north, foraging for insects. A Stellar's jay perches on the top of a fir. An Audubon's warbler hops from branch to branch in a white pine. One of my favorite little birds, conspicuous for its yellow rump and crown and the yellow patches on its breast, it makes sorties from the same tree, gobbling bugs as they pass.

A fine piece of luck for my work and my leisure to be one and the same.

For most people I know, this little room would be a prison cell or a catafalque. My movements, admittedly, are limited. I can lie on the cot, sit on the stool, pace five paces before I must turn on my heel and retrace my steps. I can, if I choose, read, type, stretch, or sleep. I can study once again the contours of the mountains, the sensuous shapes of the mesas' edges, the intricate drainages fingering out of the hills. On windy days in spring I turn my gaze upon the desert, a feast of eye on country if you like your country sparse. In early afternoon I follow the formation of dust devils through my field glasses. Their manic life and sudden death seem to me a fruitful field of inquiry when the mind bogs down in solipsism. Far off on the desert floor, where once a great inland sea bubbled, the earth rises to the dance. Scorched by sun and scoured by wind, the ancient seabed surrenders itself to points east, eventually to be washed to the Gulf in the current of the Rio Grande.

Amid a forest that burns and a desert that dances—20,000 square miles of cruel and magnificent country—I turn back, at the end of the day, to the earth beneath my feet. As May begins, wild candytuft bloom beneath the pine and fir, the first of the season's wildflowers to show their color. A relic turns up one evening in the dirt, not far from the base of my tower: a Mogollon potsherd, white with black pattern, well more than 800 years old. I am given to understand that the people once gathered in the high places and brought with them their crockery. They sacrificed their pots by smashing them to earth in hopes the sky gods would grant rain. Clearly I am not alone in my communion here with sky. Far from it. The ravens and the vultures have me beat by 200 feet, the Mogollons by most of a millennium. And who's to say the motes of dust don't feel joy, if only for a moment, as they climb up into sky and ride the transport winds?

LIKE ALL LOOKOUTS, I PURSUE diversionary measures, little games or physical routines or time-devouring hobbies that give form to the days and let me escape the holding cell of my own thoughts, particularly when those thoughts begin to circle on the metaphysics of whirling dirt. Gary Snyder practiced calligraphy and meditation. Edward Abbey pitched horseshoes with his pa on the rim of the Grand Canyon. Jack Kerouac studied the Diamond Sutra, wrote an epic letter to his mother. If I were a more dutiful son I'd do the same. Instead I shoot Frisbee golf.

The shape of my mountain invites the game. Not many peaks in the Gila offer as much flat open space as mine. On some you'd toss a Frisbee and never see it again. That can happen here on the windiest of days, so I pick my moments. My four-hole course begins and ends at the outhouse. The par five first goes over— or around—the cabin and all the way across the meadow to a big lichen-covered rock. A short par three takes me from the rock to the elevation sign next to the trail. (APACHE PEAK LOOKOUT. VISITORS WELCOME. ELEVATION 10,010 FT.) From there I embark on a difficult par four into thick trees on the south edge of the meadow, down to the tack shed by the old corral, now in disrepair from decades of neglect. Then I wing the disk back up the hill to the outhouse (par four again). Though I'm not competing against anyone, I thrill to a fine toss and cringe when the Frisbee goes astray. To this day I've never birdied all four holes in a row.

The game resembles the act of writing in that you can't help but compete with yourself: no one is watching, but you still desire perfection, even if such a thing is unattainable. In fact, the toss of a Frisbee is a bit like the writing of a sentence. Each must move along a certain line to keep the game going forward. Each can go astray, spin out of control. At times what is called for is a long, unspooling line, a toss that slices and circles and hovers in the wind, feinting one way before turning back in another, just as a sentence can move in spirals around a central idea, curving ever closer to the center, the heart, the rock. Other times you need a direct approach. Straight and crisp. A shot from short range. It helps me to think to move like this, spinning the Frisbee with the torque of my arm and then following its path through the meadow grasses, allowing my mind to play on the walk between throws, moving between noticing and imagining, memory and wish, concentration and daydream. I often find three or four rounds lead me back to the typewriter, so that the dance of the Frisbee has become a kernel of thought, thought has become a run of sentences in the mind, and the words emerge at last as ink on paper: *Time spent being a lookout isn't spent at all. Every day in a lookout is a day not subtracted from the sum of one's life.* At which point, having rendered the mood of an entire day in a derivative twenty-five-word aphorism, I pour myself a glass of bourbon on ice.

I never touch the stuff on the clock, although I can't say the same for all of my fellow lookouts. One summer a young woman from the state of Texas came our way to work the relief lookout gig on two different mountains, in which fashion she could string together ten straight days of work like the rest of us regulars, with a day and a half of that involving travel between posts. She moved between Loco Mountain, attainable by a five-mile hike, and Monument Mountain, reachable by truck on a long dirt road, and perhaps the incongruity between two different peaks tweaked a nerve. One day she got into the sauce early and began to call around, looking for a partner in conversation. Problem was, she used the main forest frequency, becoming audible to several dozen or even a hundred different people, smack in the middle of the afternoon. Firefighters, trail crews, dispatchers, other lookouts, anyone with a radio scanning "forest net," as it's sometimes called, were privy to the slurred meanderings of her search for human contact. She did not rejoin us the following season.

Once, on my days off in town, I was warming to the feel of the pool table in the Buffalo Bar—Silver City's finest purveyor of adult beverages—when my next opponent, after feeding his quarters and racking a game of eight-ball, extended his hand and said, "Buddy Nunn. Pleasure to meet you. What do you say we play for a beer?" That surname struck me, because not long before I'd spent an afternoon copying into my notebook every name penciled into the corners of my tower. Stretching back to 1939, each man or woman who'd worked the peak had followed tradition and left a name and year for posterity. Including a certain Tuffy Nunn, 1968.

"Yessir," Buddy said, when I asked if he counted the man a relative. "That there's my old man. You're right, it was sixty-eight. I was just a kid back then, but I remember riding horseback to visit a couple times. What a view from there, about the best view in all of southwest New Mexico. He only worked it one season but he sure loved it."

I asked him if he recalled anything in particular from his father's season on high. "By God," he said, fingering the feathered corners of his mustache, "the old man did have himself a time." Seems one summer evening a few friends of Tuffy's paid a visit from the world below. They packed in on mules, brought a bunch of steaks and a few cases of beer. The drinking began that night and continued into the next day. By midafternoon the party was in full swing, but the beer was gone. Tuffy got on the radio, ordered a helicopter for an urgent resupply. All was quiet on the forest, so the chopper arrived in short order. The pilot landed in the meadow. Tuffy climbed aboard.

"Down to the ranch, I presume?" the pilot said. He knew the Nunns owned a ranch on the south end of the Black Range, up White Earth Creek, nine miles away as the crow flies.

"Hell no, son," Tuffy told him. "You steer this ship to the Pine Knot Bar in Truth or Consequences, and I'll take her from there."

The pilot was just back from a tour in Vietnam. He was happy to be flying the forest with no one shooting machine guns at him. What did he give a damn where the mission led? He pointed the nose of the chopper northeast, toward the valley of the Rio Grande. Twenty minutes later they set down in the bar's parking lot, which was then, as it is now, a slightly sloping swath of gravel around a wooden building a little ways west of the river. Dust rose in a great cloud. Shards of rock flew in all directions, bounced off the sides of pickup trucks. Tuffy ducked and ran. Five minutes later he emerged from the bar with all the beer he could carry. He was gone from the party less than an hour.

"Things were a little bit looser back then," Buddy told me, his blue eyes twinkling beneath his cowboy hat. "I'm not saying the bigwigs didn't mind. They gave the old man hell all right, and they sure as shootin' didn't ask him back the next season. You'd be hauled off to jail for a stunt like that now. And the pilot—he'd never see his flyin' license again."

One way to read this story is as a picaresque example of the fact that every lookout has something he cannot live without. For Jack Kerouac it was cigarettes; two weeks into his stint on Desolation Peak, in the North Cascades in 1956, he radioed his boss and begged him to bring a tin of tobacco and a sheaf of rolling papers across Ross Lake by boat, to a trailhead five miles from his mountain. Kerouac gladly made the ten-mile round-trip hike to pick them up. For some of the rest of us, maybe our weakness is spirits, or pudding, or mystery novels. Maybe it's Major League Baseball on the AM radio. Often, nowadays, it's a cell phone.

I manage to subsist without one, though most of the lookouts I know have been packing theirs along for years. Presumably anyone calling a lookout already knows the answer to the first question a cell phone compels us to ask: *Where are you?* I do my part for social harmony face-to-face across a bar top seven months a year, which leaves me about talked out for the other five. No acquaintance of mine would intrude on my life here except under power of her own two legs; all my friends know quite well the reasons I love it, and I'm happy to share the place with them, but only in person.

That thing some people call boredom, in the correct if elusive dosage, can be a form of inoculation against itself. Once you struggle through that swamp of monotony where time bogs down in excruciating ticks from your wristwatch, it becomes possible to break through to a state of equilibrium, to reach a kind of waiting and watching that verges on what I can only call the holy. One indisputable charm of being a lookout is the sanction it offers to be shed of the social imperative of productivity, to slip away from the group hug of a digital culture enthralled with social networking, the hive mind, and efficiency defined as connectedness. I often think of a line from Aldo Leopold: "Nothing could be more salutary at this stage than a little healthy contempt for a plethora of material blessings." I want to lengthen, not shorten, my attention span, and most of the material splendors of the twenty-first century bully me in the opposite direction. The fault is mine, I'll admit. I'm too slow-witted, reluctant to evolve, constitutionally unable to get with the program. I can't afford the newest gadgets and I'm not a natural multitasker.

Whether by solitary temperament or sheer cussedness or some unholy amalgam of the two, I prefer to live, at least part of the year, out here on the edge, where worship of the material recedes and acquaintance with the natural becomes possible, and where I carry on my most important conversations through the United

States Postal Service, a month elapsing between missives from one side or the other. That seems about right for most things worth saying. For laughs I sometimes try to imagine the horror Henry David Thoreau would feel were he to fall from the sky today, the man who looked with disdain on the invention of the telegraph: "We are in great haste to construct a magnetic telegraph from Maine to Texas, but Maine and Texas, it may be, have nothing important to communicate."

In the world below I tend toward the attitude of the bemused spectator. I use landline telephones, I answer e-mail, and I've yet to renounce my wintertime access to a high-speed Internet connection. (Hypocrite!) But I can't in good conscience apologize for the fact that for a few months a year I choose not to choose anything but what I read, what I eat, and when I sleep. My interests aren't tracked, aggregated, and commodified, sold back to me in a digital feedback loop requiring no more effort than the click of my finger. Up here I'm not a six-foot-tall billboard or a member of a coveted demographic; I'm a human being, and as such I find it restorative to be in the presence of certain mysteries our species once knew in its bones, mysteries ineffable and unmediated. If I were a committed technophobe or a purist, I'd forswear even my typewriter and my AM/FM radio, with its faint and fuzzy night sounds of baseball from Phoenix or Denver, tunable some days after dark. But unlike a telephone, the radio lets me listen without demanding a response, and the typewriter's staccato music bothers no one but the birds. Even they don't appear to mind all that much. Perhaps I'm not prolific enough.

There's a seduction to solitude in a stretch of the world as we were given it, a seduction that stretches across all human cultures and all human history. It may be mocked as foolish, childish, antisocial, misanthropic, retrograde, reactionary, fuzzy-headed, and sentimental, but it exists in the human heart and will endure as long as *Homo sapiens* survives in even so much as one tribe. Out here, the kin in which that feeling thrives is among the lookouts on the wilder, south half of the Gila. This may be true in other places where our kind still exist, but it's undeniably true here: whoever works the full-time shift among us tends to return, year after year, while the relief lookouts drop away after a couple of years at most, a couple of days or weeks in some cases. Which may just prove you need a good stretch of alone to really fall in love with it. I've been here eight summers in a row, and I'm still the greenhorn on the south half of the Gila. Up on Loco Mountain, Jean has worked nine years. John has logged a decade on Cherry Mountain. Over on Monument Mountain, Hedge is now a veteran of eleven summers. And on the biggest and wildest lookout in the whole forest, Snow Peak, Razik has worked nineteen years and Sara a remarkable twenty-eight. Most of us, if we could change one thing, would either make our seasons longer or forgo days off, the longer to enjoy our state of grace and the quicker to attain it. Once you can sit on a stool for an afternoon, unmoving and unmoved by anything but light on mountains, you have become a sensei of the sedentary and need answer to no one for it, except perhaps your husband or your wife.

MAY IS RELENTLESS WITH WIND. No matter the length or sweetness of a reprieve, the wind always returns, gales to test the endurance of anyone exposed in a high place. Trees that elsewhere grow a hundred feet tall here hug the ground like shrubs, shrunken, gnarled, and twisted, as if cowering from an invisible foe. In spring, though, the foe is briefly visible: out of the desert to the south the sandstorms encroach like a thing alive and malevolent. There's no place to hide. My tower rises up like a taunt. I search my innermost self for a haven, a hideout, a moment of remembered thoughtless bliss. For long periods—minutes that feel like hours, hours that feel like days—I cower like the trees, stunned and borderline cataleptic. The guy-wire cables off the corners of the tower twirl like jump ropes, and every so often they cut a gust in half and make it scream; I startle from the cocoon where I've been hiding. I stand, look around, make sure there's no smoke to be seen. I glass the distant hills being swallowed in dust. Now and then a cloud rises off one of the open-pit copper mines thirty miles southwest, beyond the edge of the forest; immediately my eye is drawn that way. The detection of such shapes is, after all, my purpose in life. Five seconds, though, and I've ruled them out. Their color and their movement don't jibe with a smoke. Too yellow, too stationary; one minute here, the next minute gone. Early smoke is almost always white and almost always lingers.

Throughout the afternoon the visible world shrinks. Air takes on texture and color. A pine needle cluster the size of a squirrel's tail darts past the window so quickly my eyes can't follow it, although I'm sure I saw it there briefly, sixty feet above the peak and soaring like a bird. I begin to wonder if I'm hallucinating. I know for sure I'm vibrating, along with the floor and the walls of my tower; I feel as if I'm standing on a dance floor built on springs.

The radio squawks. I have to turn it up to hear it.

"Dispatch, Snow Peak."

"Snow Peak, dispatch."

"A weather update for you. I have winds of fifty-five to sixty-five, gusts to eighty, out of the southwest. Steady like that for the last half hour."

"Copy that. Thanks for the update. Dispatch clear, sixteen to twenty."

Snow Peak is usually windiest, being 800 feet taller than its nearest competitor—my peak—for tallest lookout on the Gila. My gusts reach only seventy-two miles per hour, with a steady breeze at forty-five to fifty. I try my damnedest, but—curled in the fetal position on my cot, listening to the snap and howl of the guy wires that anchor my tower in bedrock—I can't find much consolation in knowing it could be worse.

Some nights 6:00 p.m. comes and I find myself reluctant to leave my room with a view. This time of year the hammered mesa tops glow pale blond in the low-angled light, and glades of aspen can be seen greening up here and there in a kind of mosaic where the old McKnight Fire burned, each dense cluster at a slightly different pace—one vegetation type, a dozen different shades of green. Under the evening sun they have a fluffy look to them. But on windy days such as this I bail a few minutes early, pack a sandwich and a bottle of water for the trail. The roar has driven me to a deep disquiet. I need to get off the mountain.

Of all my evening destinations, the pond may be my favorite. Tucked out of the prevailing winds in a kind of alcove, a mile and a half below the peak, it holds water most of the year and is visited by elk, deer, bear, and turkeys; their tracks are often visible in the mud. Alice loves it here too. Usually she opts for a soak in the pond's fetid water—unless she finds a pile of bear scat to flop in first, in which case she rolls around on her back, kicking her legs in the air as she smears herself in the stink. Few things make her happier than finding something dead or dungish in which to roll. The reek of it is a kind of talisman for her. It says, *Beware, all ye who travel here: I am as mean and nasty as I smell.*

Around the pond this time of year, the spearlike leaves of *Iris missouriensis*—commonly known as the Western blue flag—poke through the mud, straining toward the sunlight. The most drought-resistant of wild irises, it sings the song of springtime in the high country. Out of the hundreds of flowers forming pregnant buds, precisely one has burst, exposing three delicate petals traced with threadlike purple veins. I sit cross-legged next to it and rest awhile. High on the cliffs above me I can hear the wind roar, but here in my watery bower all is calm. For the first time all day I'm able to hear myself think, were I to have a thought. The wind has left me hollowed out, though. I haven't spoken a word since I called in the weather ten hours ago.

HUMAN CONTACT HERE IS the more cherished for its rarity, and my favorite encounters have been with that peculiar subspecies known as the thru-hiker. Their aim: 3,000 miles on foot in five months, a hike along the Continental Divide Trail (CDT) from the Mexican border to Glacier National Park before the snow flies in October. Twenty miles a day, every day. I know them instantly by their fancy walking sticks, their sunburnt skin and general air of dinginess, and the scratches on their shins if they're wearing shorts, which come from having bushwhacked through the thorny brush to my south. By the time they arrive on my peak they've walked a hundred miles on a shortcut from the Mexican border, headed for a junction with the Continental Divide just northwest of me. The actual divide runs north out of the bootheel of New Mexico, but that way lies desert, road walking, and a land of dangerously scarce water. By heading straight north from Columbus, New Mexico, site of Pancho Villa's 1916 raid across the border, the thru-hikers save perhaps forty miles—two days of walking—and can reach the true divide within a day of seeing me. There's no such thing as a day off in their lexicon, only what they call a "zero day." For these people walking is joy, not work, yet daily mileage remains an axiom of progress. Their resupply points have all been planned in advance. Falling off the pace can mean going hungry. Sometimes friends or family back home will mail them boxes of good trail food, which await at rural post offices spaced between seven and ten days of walking apart. Others take side trips or hitch rides into little towns near the divide, restocking on gas-station food: Doritos, beef jerky.

I find these folks unfailingly gracious and cheerful, with their lightweight equipment, their hard legs and big smiles. They're a self-selected bunch, at ease in the out-of-doors, but for people who've been walking close to a marathon every day they appear almost goofily invigorated. Within moments of their arrival they shed their packs and ask to climb the tower. They're a week into their five-month journey and want to see where they've been and where they're headed. I'm always happy to show them what I can: a couple-hundred-mile stretch of their walk, from the Mexican border to the country up beyond the Middle Fork of the Gila River. It's a land of stark vistas and rough country they've traversed—dry, wind-scoured, humming with ancient mystery, dotted with hidden petroglyphs.

With most CDT hikers I find a sort of mutual envy. I admire their courage and stamina and sheer gumption, their tolerance for every sort of backwoods discomfort. They admire my solitude, my view, the countercultural weirdness of my job. Each of us has a taste for wild country. I sit above it, letting it come to me in color and shadow and light. They pass through it on calloused feet, stopping to make camp each night in a strange new slice of the world. Their challenges are profoundly physical: rattlesnakes, biting flies, blisters, screaming Achilles tendons, thunderstorms, extremes of heat and cold, all the pleasures and pitfalls of life outdoors on the move. Mine are existential: time, space, the sweep of geologic epochs written on the view out my windows, which remind me I'm but a mote in the grand saga of Earth's history.

When they leave, I often wish I could go a little ways with them.

The CDT is one of the three long north–south walks in America—the others being the Appalachian Trail and the Pacific Crest Trail—and by far the most difficult. For those who undertake these cross-country treks, nicknames are virtually obligatory. The first to greet me this year are a pair who go by the names Reno and Slouch. Reno is a slim woman, dark-eyed and olive-skinned; Slouch is a tall, pale, sunburned Brit with a growth of bright red beard. They are amiable visitors, grateful for fresh water, happy to step inside the cabin and drink a cup of coffee from my French press. I offer a snack of crackers and grapes. Reno reciprocates with some hand-rolled cigarettes. Slouch says he's from Southampton, in the south of England, and because he has a bookish look I offer him a recent copy of the *London Review of Books*, for which he seems both shyly grateful and a little bit stunned. "If you had told me, before I began this queer odyssey, that I'd meet someone in the middle of the woods of New Mexico with a subscription to the *LRB*, I'd have told you you were stark raving mad," he says.

Having seen no one in days, they are eager to tell me the story of how they were shadowed by a Border Patrol agent on the very first day of their walk. He came crawling out of the brush in the desert behind them, just as two guys in a Border Patrol vehicle pulled up ahead, all of them armed with pistols and wearing wraparound shades. The agents took one good look at their quarry and realized their error; they expressed bafflement that anyone in right mind would take a stroll to Canada for pleasure. Earlier Reno had taken a few pictures of the Mexican town of Palomas from the American side of the border. This aroused the suspicions of the authorities, and she and Slouch were forced to show papers despite not having crossed into Mexico. Reno worried about her passport smelling of marijuana, having kept the two together in the same little bag. She hadn't expected an ID check in the lightly inhabited desert of the American Southwest. Luckily, the border agent did not have a first-class sniffer, or her journey might have ended before it began.

"We're not going to see any more of them, are we?" she asks, and I tell her that from here on out they should be in the clear.

A few days later a CDT hiker appears at 9:00 a.m., another red-bearded fellow, this time named Dave. I know what he's up to before he even opens his mouth, partly because of the week-old beard, which is de rigueur for male thru-hikers. Dave camped the previous night at Wright's Saddle, not far from my truck. Like most of his kind, his first question is: *Have you seen any others?* They're always interested in the progress of their fellows, whom I've seen and when. Many announce their intentions over the winter on Internet forums, asking for advice on routes and logistics, so a certain camaraderie exists among people who've never met but are undertaking the same grueling journey. When I mention Reno and Slouch, he smiles and says, "I bet I can catch them within the week. One person's always quicker than two."

Dave is the most philosophical of the thru-hikers I've talked with. He tells me how he worked on an organic farm in Hawaii with a woman he loved deeply until their relationship curdled for reasons he did not understand. Having been with her some time, having lived in a little hut with her, built a life with her, the sudden loss of it all was like a punch to the gut. He set out on the Pacific Crest Trail last year seeking—he knew not what, exactly. A connection with wilderness. A means of forgetting. A means of remembering. A means of making peace with pain. The metronomic quality of walking allowing him to feel or not feel as he wished. The attainment of a state like a trance. Coming down off Fuller Ridge in California, over a landscape charred by a wildfire, he felt himself overcome. He leaned against a dead tree, gripped it for balance. The tears began. His nose ran. He felt every emotion at once: sadness, anger, joy, horror, love, hope, despair; every cell in his body quivered in anguish and ecstasy. He both lost himself and felt himself more intensely than at any other moment in his life. And in the aftermath of this moment—this vision, this transcendence, this coming into the world, this death and rebirth—he felt emptied and calm. Ready to begin his life anew. Open to mystery and beauty, reconciled in some ineffable way to loss.

"It was as if I had tapped the source of some divinity," he says. "Our language does not contain the words to describe it. Perhaps a better way to say it is that some divinity coursed through me; I became the conduit. I was holy and I was nothing and I was inseparable from the world around me, no boundary marking the charred landscape and my self and where they met and merged. I've never experienced anything like it without the aid of psychedelics. Now I can't imagine doing anything else but walking in the wilderness. I've shed every worldly and material ambition; I have no desire for anything but that complex register of feelings I can only get on a months-long walk alone in the desert and the mountains. Boredom, ecstasy, hunger, thirst, blisters, sunburn, mania, longing—the whole crazy carnival."

And then, having bared his soul to a stranger in the middle of the woods, having queried me about good sources of water in the miles ahead, he shakes my hand, wishes me well, marches across the meadow, and disappears into the trees.

.　　.　　.

NOT EVERYONE WHO WANDERS into this corner of the forest is a thru-hiker—far from it. Most folks are simply out for a day hike. Then again, there are others embarked on even stranger and less definable journeys than the quest to trace the CDT. One evening I meet a peculiar woodsman of the American West, a roving packer by the name of Dane, from Montana. I am on my customary evening stroll when I hear a bell and a voice. The bell turns out to be strung around the neck of one of Dane's horses. "In case a bear comes around and she runs off, I can find her again," he explains. I sit with him at his camp next to the pond. He tells me he moved from Michigan to Montana eighteen years ago, bought a horse and a book about breaking horses, and now rides through the West 300 days a year with his saddle horse and two more for packing. His girlfriend mails him food so he can ride pretty much freely for months on end, detouring to the nearest post office when necessary. He goes home a couple of times a year but always strikes out again, for Wyoming, Idaho, Nevada, Utah, Colorado, New Mexico, Arizona, California, Oregon, and Washington. This year he wintered on the Rio Grande, thirty miles east of here, while one of his horses recovered from an illness. Now that she's healthy they're back on the move.

He tells me that once, while riding south on the Pacific Crest Trail, he decided against following the trail through the desert to the Mexican border and turned instead for L.A. There he rode his horse down Hollywood Boulevard and ended up, eventually, in a black neighborhood, staying with a young lady whose neighbors didn't take kindly to a white interloper with three horses shacking up with one of their own. He was pulled into an alley, a pistol stuck to his ribs, and told in no uncertain terms never to show his face there again. "But hey, women love horses," he says. "Best way to pick up girls is to ride through a city on a horse."

About the fifth time I ask the question *Why do you ride?*, he finally decides to answer it directly, instead of with an anecdote: "I get antsy if I stay in one place more than a couple of months." His stories all have a mythopoetic ring about them. He tells of riding out of the redwoods onto a ridge overlooking San Francisco and being so stunned by the beauty of the night lights of the city, the bridges, the bay, that he simply had to find a way to cross the Golden Gate on horseback. He petitioned the city but was turned down: no stock allowed, a remnant law meant to deter Basque sheepherders back in the day. He decided his best chance was to cross in the middle of the night but he didn't count on the surveillance cameras—there to warn of potential suicides before they jumped—which tipped off the cops to his presence. He was a quarter of the way across when he was pinched by coppers from both directions. They didn't take kindly to his request, since he had come that far, to turn a blind eye and let him cross.

Weirdly, he's no minimalist. He travels with portable solar panels, 12-volt batteries, a laptop, a Game Boy, a cell phone, a shortwave radio, and an infrared burglar alarm that he sets each night around his camp. In L.A. he was nicknamed the Electric Cowboy for all of his gadgets. Before I leave to hike back to the peak he tells me he's about to retire to his blowup mattress and the NBA playoffs on the radio. "If you're ever in Montana," he says, "don't bother to look me up. I probably won't be home. But maybe we'll meet again in the woods some day."

A WEEK PASSES WITHOUT SO much as one hiker stumbling into the meadow. Amid an exalted solitude I become an aristocrat of sky, an aristocrat of time and space. On the FM radio I hear news of war, greed, corruption, hypocrisy—same as it ever was. I do not listen long. I prefer the silence, the sloth, the sweet stupefactions of landscape worship.

Every so often my stripping away of need and worry leaves me eager to be shed of even more. Four walls and a roof begin to feel like an encumbrance. The lengthening days and the coming of the full moon provide the excuse I need to leave behind what little I have in the way of modern comforts, if only for an evening. Even a mountain deserves a night alone now and then.

I stuff my pack with a sleeping bag, a bar of chocolate, a half-pint of whisky, two liters of water, a packet of camping matches. Alice eyes me intently, moans a little, her eyebrows twitching with curiosity. Something's up and she smells it. *You and me, Spookeen*, I say to her. *We're going fishing*.

Amid the myriad ways I'm lucky here, I count chief among them my freedom to leave the two-way radio behind at quitting time and hike with my dog and fly rod down to a stretch of trout water seven miles from the lookout and as far from the nearest road. If I hurry down the trail I can play on the creek for an hour this time of year, then hike partway back by moonlight. As long as I'm on the radio by 9 a.m., no one need know my night movements but the owls and the trout. I like the walk as much as I like the fishing, which makes me unconventional among fly fishermen, who prefer their trout water, generally speaking, a heck of a lot closer to the truck.

Two miles down the trail, in a little meadow a hundred yards below me, I see something large and dark rear up on its hind legs, its front paws curled in a tuck before its torso—a bear with its snout in the air. The snout points toward the tinkle of Alice's collar where she walks a little ways ahead of me. I stop and crouch behind the trunk of a big Douglas fir, spy for a moment on the bear, my veins flooding with adrenaline. When the bear shows no sign of leaving, I clap my hands several times. The bear wheels, drops to all fours, saunters away through the meadow. Alice looks at me queerly, none the wiser. I'm not sure she'd agree that part of the thrill of the walk involves moving through country where neither one of us is the top link in the food chain. Black bear country isn't nearly as spooky as grizzly country, but black bears have been known to maul a human now and then—generally around campgrounds where they've become habituated to humans offering food. Still, their presence keeps my senses preternaturally alert to sounds, movements, colors. At other times of the year, in other places, I'm a man with a debit card, a driver's license, a Social Security number—a quasi-functioning member of the rat race. Out here I'm a biped with tender haunches and a peculiar smell, too slow to outrun a large predator.

Along the way to the creek, New Mexico locust perfume the fire-scarred ridges with their sweet pink flower clusters, and once I reach the headwaters, bluebells demurely show their drooping blooms, accented here and there by orange and yellow columbine. Down and down we go into a canyon bracketed by orange and pink cliffs and weird squat hoodoo rocks, through groves of aspen and a scattering of huge Douglas fir. The trail meanders back and forth across the creek. Mostly its waters are small enough to jump from bank to bank without wetting my feet, though the farther down I go the more they're fed by hidden springs, until I'm forced to hopscotch on exposed rocks. Below a twenty-foot waterfall, which acts as a barrier to the upstream movement of the fish, I shed my pack and assemble my rod and reel, tie a bead head woolly bugger on the end of my leader. I call Alice over, give her the basic commands: *Sit. Lie down. Stay. Be good.* The stream is challenging enough for a novice like me. I don't need the added worry of hooking her in the ear.

What this kind of fly-fishing lacks in poetry and grandeur, it makes up for in finesse and stealth. The narrowness of the creek limits my back cast, as does the overhanging vegetation—for much of its length the stream resembles a tunnel. I'm lucky to find a hole I can hit from fifteen yards away. I sneak up behind boulders on all fours, fish holes hip-high above me from below, cast from all sorts of awkward postures—crouching, kneeling, sitting. I throw a fly into each hole three or four times, and if nothing rises I move on. The fish I do catch are small: mostly eight or nine inches in length, with an occasional thirteen-incher. Some of these trout are mongrels, many of them hybrid cutbows that show both rainbow and cutthroat characteristics. Their colors run the gamut from brownish-gray to green and gold and pink, shading into orange and red, some of them with big black spots on their sides and a telltale slash of orange on either side of the lower jaw. After the McKnight Fire in 1951, ash runoff killed most of the native trout in this stream, or so it was presumed—no one can say for sure—and later the state Game and Fish Department stocked it with various non-natives, dumping whatever was at hand. Before then, the stream was home to the state fish of New Mexico, the Rio Grande cutthroat (*Oncorhynchus clarki virginalis*), now a threatened species reduced to less than 10 percent of its historic range, which once spread across 6,600 miles of mountain streams that funnel their waters to the Rio Grande.

According to scholars, the Rio Grande cutthroat appears in the first written mention of a North American trout by Europeans. In 1541, Pedro de Castañeda de Najera, a member of the Coronado expedition, noted "a little stream which abounds in excellent trout," likely Glorieta Creek, southeast of modern Santa Fe. Over the past 150 years, mining, logging, road building, cattle grazing, fire suppression, and the stocking of non-native species have destroyed the fish in vast reaches of its range. Increasingly isolated populations remain, most of them in northern New Mexico and southern Colorado, cut off from intermingling with their kind in other streams and therefore susceptible to genetic stagnation. Rising water temperatures, as a result of global warming, may also imperil their long-term survival. Government officials have so far denied efforts to list the fish as an endangered species—mainly, they admit, because they don't have the money for a recovery program.

Some of the fish I catch show the red coloring and huge black spots of native Rio Grande cutts, but intermixed with the rainbow's pink stripe; some of them lack the cutthroat slash mark on the lower jaw. Others show signs of interbreeding with Yellowstone cutts, another non-native once let loose in this creek. The higher in the headwaters I catch them, the more they look pure Rio Grande. Though their genetics are a jumble, state law permits catch-and-release only with a barbless hook—an admission that though no one quite knows what these fish are on a case-by-case basis without a complicated DNA test, nearly pure Rio Grande cutthroat still populate certain stretches, and, being rarer by the year, they ought to stay here. The fish I land I quickly place back in the water, holding them in my hand just long enough to whisper a heartfelt mantra: *Live long and propagate, my friend.* They may not, in fact, live long; the state Department of Game and Fish is contemplating, among other options, poisoning the stream, ridding it of every last fish, native or non-native, and restocking it with pure Rio Grande cutthroats. Whether it's a good idea to replace these fish, mongrel though their heritage is, with a mix of wild and hatchery-raised trout of purer genetics—that's a deeply contested question. Poisoning the fish would poison insects and invertebrates too, though some studies show the insects rebound within a couple of years.

Fishing would have to be barred while the reintroduced fish took hold, annoying the sporty types who care less about identifying exactly what they catch than they do about the fact they managed to catch it. I've seen men at public meetings on the subject foam at the mouth in anger at the fisheries biologists; I've heard state Game and Fish officials bad-mouth the Endangered Species Act for being an impediment to selling sport licenses. I tend to trust the biologists over the bureaucrats, and they tell me the best chance to have a thriving trout population in the stream for the long run is to populate it with fish whose genes evolved over millennia to cope with conditions on site—pure Rio Grande cutthroats, in other words. That places me in the camp of the poison advocates, a position that doesn't exactly give me warm fuzzies.

At dark I return to the dog, let her sniff my hands, which excites her when they smell of fish; I like to pretend she shares in the thrill of my success. I break down my rod and strap on my pack. The moon has risen above the canyon rim to the east, cool and white as bone china. Our way back is bathed in a bluish light. Three miles below the peak, a little before midnight, we stop in an open saddle at the head of two canyons. I gather some wood, light a fire, spread my bag on a soft spot in the grass. Alice snuffles through the Gambel oak, crunching in the fallen leaves of last year, sniffing for a sign of something to chase. I eat a little chocolate, drink a little whisky. The Big Dipper tips into view, encircled by the treetops around the meadow's edge, and below it Draco the Dragon's tail curves. For a while I whistle back and forth with a whippoorwill, trying to call it in close for a look, but it tires of my effort at mimicry and drifts off into silence. Now and then the wind comes up, blows some smoke off the embers of my fire, scents my dreams of trout and bear.

In the morning I douse my fire with all my remaining water

and toss dirt on the sodden ash to make sure the coals are dead out. I do not want to be known forever as the lookout who burned down his own mountain.

IT IS THE SECOND WEEK of May, and we haven't seen a spot of moisture since a snowstorm in mid-March dropped eight inches on the crest. I know because I was in it.

My friend Black Larry and I had planned a four-day backpack into the heart of the Black Range, a pre-fire-season look at the high country. We'd chosen a route entering the mountains from the east amid foothills of around 6,500 feet above sea level, climbing ten miles to the crest around 9,000 feet, following the crest north for ten miles to above 10,000 feet, and looping back to our starting point—more than thirty miles in all.

Black Larry has been a friend for years, since the moment he walked into the bar where I worked and we discovered our mutual love of baseball and backpacking—as well as the fact that we shared an alma mater, the University of Montana. (Go Griz!) He'd earned his nickname from an unfortunate tendency to harm himself while mountain biking, having broken a hip, an arm, and a finger in various accidents. Soon after our meeting we undertook to explore every corner of the Black Range on multiday trips. He figured mountain hiking would be less likely to end in disaster than mountain biking. On our first outing we lost the trail and bushwhacked for eight hours only to find ourselves back where we started, minus Black Larry's hiking poles and about a pint of blood apiece. On every subsequent outing we've encountered bad weather: rain, lightning, hail storms. So it came as no surprise when, on the second day of this trip, after a punishing first-day climb up the South Fork of North Dry Creek, we woke to strange clouds and temperatures of 12 degrees Fahrenheit. Being the pyromantic among the two of us, I got a fire going to boil some coffee and heat our frozen fingers and toes. When I shouted at Black Larry to get out of his sleeping bag and get moving, he unzipped his tent flap and thrust his middle finger in the direction of my voice.

"Have some hot coffee," I said. "You'll feel better about life."

"I'm trying to feel okay about the prospect of freezing to death."

I laughed, added some wood to the fire. The prospect of death was one of the unspoken reasons we took these trips, and we both knew it. He owned a historic brick home back in town, had a smart Scandinavian beauty for a wife and two grown children making their way in the world, one a journalist, the other a lawyer. He worked fifteen hours a week as a consultant for a Fortune 500 company at some obscene hourly rate. He was on pace to retire well before the end of his fifties. (His motto is that of all highly paid consultants: "If you can't be part of the solution, there's good money to be made in prolonging the problem.") His daily life was structured for maximum ease and comfort, in other words, and while this had been the plan all along, on some level it troubled him. His broken bones were merely the most vivid evidence that he needed a brush with danger now and then, to keep his senses alert, stave off the onset of senility, remind himself he was, at base, just an animal—a highly evolved animal. It's not as if either one of us loathes our domestic life. We love our wives, good wine, sports on television, dinner in a nice restaurant—especially our wives. It's more a matter of achieving some measure of balance, some substantial contact with that part of ourselves that relishes a campfire under a sky berserk with stars, forty miles from the nearest social worker, completely reliant on our own dexterity.

Within an hour of our breaking camp that morning, snow had

begun to fall. We'd structured our trip with this possibility in mind, figuring we could reach the old abandoned lookout on Nana's Peak in a pinch. Too far from a decent road to get people in and out with any ease, the lookout had been shuttered decades earlier. From the spot where we'd camped that first morning, Nana's Peak was another nine miles, and we had no choice but to press on through the storm. We hiked beneath a gorgeous forest of spruce and fir and through the scars of old burns where blown-down aspen lay across the trail by the dozen. Snow collected on our hats and melted from the heat of our heads, refreezing into grotesque ice sculptures in the cold. We couldn't see fifty yards in front of us, the snow was falling so hard, and almost every step was fraught with the possibility of disaster—a twisted knee, a sprained ankle. I hadn't felt so giddy in months.

We arrived on Nana's Peak after five hours of tortuous walking and found the old cabin intact, though filthy with rat shit. Some good soul had left enough wood in the box to get a fire going, and Black Larry swept away the rodent droppings while I tended the hearth. Once we'd warmed ourselves and cleared away the worst of the filth, we found an old metal bucket and filled it with snow to melt on the stove, replenishing our water supply. All night the blizzard raged, but inside the cabin we were warm and dry. We found a deck of playing cards, a couple of candles, a pen and paper, and we settled in for cribbage at the crude table, keeping score the old-fashioned way.

"Why don't we make it interesting and play for whisky?" I said. "Winner gets a shot of the other person's stash."

I had taught him how to play myself and figured I could school him as I usually did, capturing more than my share of bourbon in the bargain. I was wrong. He kept getting hands; I kept muttering, "This ain't a hand, this is a foot," and before long he was gibbering happily about his prowess at cards and my whisky supply had shrunk to next to nothing, though I'd barely touched it.

"You have learned well, grasshopper," I said.

"Grasshopper like whisky," Black Larry said.

In the morning we woke to a world resplendent with light. Not a cloud could be seen in the sky, and everything glittered under the sun. We were momentarily blinded when we stepped outside the cabin. By noon the snow was melting off the roof, so we set the battered bucket under the eaves to catch the fresh water. The old tower beckoned, despite its wooden steps having blown away near the top. We carefully climbed its skeletal remains to look out on the Black Range in all its winter majesty. The scars of big burns stitched a patchwork of forest types, with the vegetation everywhere in some state of recovery from big fires—the Divide Fire, the Pigeon Fire, the Bonner Fire, the Seco Fire, the Granite Fire, each of them covering thousands if not tens of thousands of acres over the past twenty years. From where we stood there wasn't a road within ten miles in any direction, and not a paved one within thirty. We were alone above 10,000 feet, in the heart of the heart of the Aldo Leopold Wilderness, fifty miles of mountains stretching to our north and the same to our south.

"I'll bet we're the first people to make it here this year," Black Larry said.

That night he graciously shared his whisky, my supply having been decimated, and I found an old scrap of paper I tacked to the wall, after writing on it the following message: "On this peak and on this rock Black Larry cleaned my clock at cribbage on a snowy night in March—the South End Sentry." Meanwhile, Black Larry had been doing some writing of his own. "Want to hear Black Larry's Rules for Black Range Travel?" he asked.

"You're some kind of guru now?"

"Calm yourself and listen, son. You might learn a few things from a Black Range hand like myself."

1. Do not speak of Black Range travel in the language of conquest; the Black Range will make you its bitch.
2. Pack a minimum of 4 ounces of whisky per person, per night, and do not gamble with it.
3. Avoid the high country before May 1, unless you're packing snowshoes.
4. Use maps as rough general reference only; trails on maps may not exist on the ground.
5. Add 50 percent to mileage on trail signs—all of them lie.
6. Layover days are highly advised for anyone over the age of thirty-five.
7. When in doubt as to your route, refrain from bowling up prematurely.
8. Make friends with people in high places; they have Forest Service keys to locked gates and cabins.
9. Always stow cold beer in the truck for the end of the trip.

"One more and I can call them commandments," Black Larry said.

We considered several additional maxims, each of them with merit, before we settled on one that spoke to our immediate situation but also had a timeless quality: *Thou shalt not complain of snow or rain in a land where they're seen infrequently.*

.   .   .

Since then, two months of sun and wind have depleted
fuel moistures. On the morning fire-weather forecast, fire danger
moves for the first time from High to Very High. The same afternoon, as if on cue, I see a bloom in the shape of a mushroom,
thirty miles west of me, drifting up over one of the ridges running
north of Cherry Mountain. It settles a bit, drifts, blooms again.
Unmistakable. Smoke.

I turn to the binoculars, focus in, confirm the judgment of the
naked eye. I hang the binoculars back on their hook, circle the
firefinder cabinet, spin the ring with the flat of my palm. I squat
for a look through the peephole, adjust it to settle the crosshairs at
the base of the smoke. It's so far away I need a moment to work
up an accurate azimuth, but eventually I get it: 273 degrees, 45
minutes.

Next I drop the big map on its hinges from the ceiling. I wrap
a length of string around the nail driven into the map at my location, run it out over the compass rosette at 273 degrees and three-
quarters. North of Cherry Mountain the string crosses a paved
highway, right over a spot called Grandview Promontory. Plau-
sible, especially in the absence of lightning. Whether out of malice
or carelessness, there can be only one cause—*Homo sapiens* and
he rarely strays far from a road.

Normally my next move would be to call Cherry Mountain
for a cross, but no one's home there. In fact, with the fire on the
far side of a ridge from here, a mere six miles from Cherry Mountain and five times that far from me, it seems likely that if Cherry
Mountain were manned I'd have heard about the smoke fifteen,
maybe twenty minutes ago merely by having my radio set to scan.
Rumor has it that a paperwork snafu has delayed the rehire of
my friend John, who should have been on duty for weeks by now.
Until the bureaucratic cluster is untangled, the district has been

sending its fire prevention officer up the mountain most days for a look around. Today, for reasons unknown, he failed to show. Maybe he had the day off and no one volunteered to spell him. No matter now. I'm on my own—and off my turf.

It's 5:00 p.m., so the angle of the light doesn't help. The entire landscape to my west appears one-dimensional. I could take a good long while, study hard through the binoculars, try to map each faintly discernible ridge, maybe pinpoint the smoke within a quarter mile. But I figure it's better to report now and err slightly with a location, since it will take crews as much as an hour to arrive on scene—an hour they can use to get rolling and I can use to work up a more precise legal, if indeed my first hunch proves mistaken.

I offer up my knowledge to the dispatcher, make sure she knows that my legal, without a cross from Cherry Mountain, remains tentative. Within minutes an engine is on its way. By the time it reaches Grandview Promontory, I've revised my initial report to put the fire three miles farther west, up Cow Camp Canyon: thus, the Cow Camp Fire. There the crew finds my smoke, arriving on scene, according to my notes, fifty minutes after my first call to dispatch—pretty good time on a winding mountain road. They immediately call for backup: another engine, a hot-shot crew. Fire size is modest, two to three acres, burning on a flat aspect in grass and ponderosa pine, active on all sides with flame lengths of two to eight inches. Spread potential is moderate. Cause appears to be human: a truck is parked in the center of the burn.

On a fire this far away, I can be of little help beyond ringing the alarm bell. I'm too distant to offer radio relays or give the crew any intelligence about fire behavior. All I can do is sit in my tower and watch the character of the smoke for hints on the crew's prog-ress. For an hour or so there's little change. The smoke puffs, rises, disperses, disappears, puffs again. I see no sign of real growth.

With a force of more than thirty hard at work scratching a line around it, the IC calls the fire contained just after dark. As usual, mop-up duties will remain into the following day, but this fire, like most others, will barely make a mark on a landscape so vast.

SWEET, EXPANSIVE DAYS OF BIRDSONG and sunshine string together, one after another. Through the open tower windows I hear the call of the hermit thrush, one of the most gorgeous sounds in all of nature, a mellifluous warble beginning on a long, clear note. Dark-eyed juncos hop along the ground, searching for seeds among the grass and pine litter. All is quiet on the radio. Not a single fire burns in southwest New Mexico. I swim languidly in the waters of solitude, unwilling to rouse myself to anything but the most basic of labors. Brush teeth. Piss in meadow. Boil water for coffee. Observe clouds. Note greening of Gambel oak. Reread old notebooks for what this date has offered in other seasons, such as this, from three years ago:

> I wake to three inches of snow and a world of frozen silence. Fog cups the mountaintop, shrinking the visible world to a couple hundred yards in all directions. I am socked in most of the day. Not until late afternoon do the clouds break and reveal a world cleansed by rain, with views to snowy Sierra Blanca in the east and the Magdalenas, also snowy, over the flanks of the San Mateos. Showers trail over the Burros like a curtain rustled by wind. Everywhere water drips, snow falls from trees; ice dislodges in big chunks, clangs through the steps to the tower. I nap, eat, nap, drink coffee, read, nap again. Three naps in one day—an all-time record. Indolence in my tower at thirteen bucks an hour—all glory and honor to the American taxpayer, who keeps me hard at work!

How did I come upon this aptitude for idleness? I blame it on the injurious effects of my Midwestern youth. At age six I learned the logistics of cleaning manure from the family hog barns. Around the same time I joined with my brother in plucking rocks from plowed fields and pulling weeds by hand from neat rows of soybeans. Manicured fields and well-kept barns—the whole right-angled geometry of Midwestern grain farming and its attendant animal husbandry—eventually became synonymous in my mind with a kind of pointless feudal labor that condemned its practitioners to penury or government handouts. At twelve, after the bankers invited us to leave the farm, I took on odd jobs in town—mowing lawns, raking leaves, shoveling snow, gathering aluminum cans to sell at the recycling plant. At fourteen I began a short-lived career in the grocery trade, bagging foodstuffs and mopping spills in the aisles, occasionally filching a box of Little Debbie snack cakes in compensation for a paltry wage. At fifteen I learned to fry donuts in our small-town bakery, 3 a.m. to 8 a.m., six days a week, a job I held until the day I left for college. To pay tuition I painted houses, baked bread, unloaded package trailers at UPS in the middle of the night. I tended bar. I dabbled in the janitorial arts, cleaning the University of Montana fieldhouse after basketball games and circuses. I spent a summer as the least intimidating bouncer in the history of Al's & Vic's Bar in Missoula. I baked more bread.

Undergraduate degree in hand at last, I ascended to the most rarefied realms of American journalism, handing out faxes and replacing empty water coolers for reporters at the *Wall Street Journal*. My tenacity and work ethic established, I was promoted to copyediting the Leisure & Arts page, a job I held for three years. I was anonymous, efficient, watchful, and discreet. Four days a week an unblemished page was shipped electronically to seven-

teen printing plants across the country, and the following morning nearly 2 million readers held the fruits of my labor in their
hands. At first I resented the lack of attention paid to my mastery
of English grammar and the intricacies of the house style book.
Not once did I receive a letter from an armchair grammarian in
Terre Haute or Pocatella, one of those retired English teachers
who scour the daily paper with a red pen in hand, searching for
evidence of American decline in the form of a split infinitive. Nor
did my immediate superiors mention, even in passing, that I did
my job diligently and well. Over time I began to take delight in
this peculiar feature of my job—that my success was measured by
how rarely people noticed what I did. I was barely noticed at all.

The essentials of my current line of work—anonymity, discre-
tion, watchfulness—are not so different from those demanded of
a copy editor, minus the need for efficiency. The lookout life fell
into my lap, no effort required. It came to me, in fact, on vacation—seemed like one long vacation itself. I surveyed my past and
saw only blind striving; I played out my future and saw an abyss:
day after day, the guillotine of an evening deadline, stretching into
the murky distance. I looked long into the abyss and I jumped.
This is where I landed. How could I refuse such a sweet summer
sinecure? That, at its essence, is the story of my talent for sloth. I
tried it for the first time in my life. I liked it. The plague of Midwestern Catholic guilt on my conscience notwithstanding, I often
feel I could work this mountain as long as I walk upright on earth.

But I won't. I know I won't. Other priorities must be accounted
for, among them Martha's long-term career goals—she'd like to
be a nurse practitioner someday, and the schooling for that will
take us out of state. She will be here tomorrow, her first visit of
the season. Her arrival is the impetus I need to rouse myself from
squalor. In the evening I sweep the outhouse, heat three pans of

water for a bath, haul the tin tub onto the porch. A rudimentary bath, the water maybe two inches deep, but satisfying: most of the cleaning action results from splashing water on myself. There will be fresh occasions to let myself go completely feral and to wallow in my own glorious stink, but this is not one of them, not when Martha is driving forty miles and hiking five more to see me. I emerge from the tub feeling cleansed, rejuvenated, though perhaps I should sport a temporary tattoo in a sleek sans serif font as I shiver in the breeze on the porch: Objects Between Legs Are Larger Than They Appear.

A COUPLE OF TIMES a season a day hiker works up the nerve to ask a personal question, and the exchange usually goes something like this:

"So, son, just curious—you're not married, are you?"

"In winter I'm married. In summer I have a girlfriend who pays an occasional visit. Lucky for me, they're the same woman. Simplifies matters. Keeps me out of trouble."

The hiker chuckles uncertainly, trying to wrap his mind around my riddle.

I met Martha in the spring of 2003. After my first half-season in the lookout, I returned to New York with the notion that I could make a living as a writer. This notion proved false, so I cashed out my *Wall Street Journal* 401(k), paid the punitive tax penalty, and lived for five months on what was left. By the time March rolled around, I was selling my books for cash at the Strand bookstore and counting the days till I returned to New Mexico—this time, I thought, for good.

Silence, cunning, exile: I would make that mantra my own.

There was a bar down the street from my apartment in Queens where I liked to sit alone and drink a couple of beers after a day's work at the typewriter. I didn't go there for conversation or companionship, though I always mimed pleasantries with the bartenders, who'd come to know me as well as circumspect people separated by a bar allow themselves to be known in a city of 8 million people, which is to say hardly at all. I went there to watch, to eavesdrop, to slip the cage of the self for a while. Since I'd begun my career as a scribe-for-hire, I hadn't missed office life at all, but reading and writing and thinking all day did have the potential to make me socially strange, and I knew it wasn't advisable to shun human contact altogether. Not just yet, anyway: I'd be alone soon enough. In the meantime, watching and listening allowed me to keep my equilibrium, avoid going around mumbling to myself like so many other lost souls in the city.

I took a seat at the end of the bar on the only empty stool. I'd brought a book, having forgotten it was Friday and the bar would be loud with music and talk, too loud to read. I ordered a beer, pulled out a cigarette, fished in my pocket for a lighter, without success. As a purely recreational smoker, I often left home without one. The woman next to me was smoking, so I asked if I could trouble her for a light.

"No trouble at all," she said. One glance revealed her as painfully beautiful, with dark brown eyes half hidden by stylish glasses, and an easy, slightly crooked smile that made me blush and turn away. I stared into my beer, unable to think of a thing to say. She let me off the hook by leaning over and whispering, "See that guy?" She tilted her head toward a voluble, intoxicated man who gestured and held court in a booming voice at the corner of the bar. I nodded. "He latched on to my roommate an hour ago and won't let her go. I think he might be the town drunk." She paused, looked over at him gesticulating, turned back to me. "Not so good in a town this big."

For the next three hours we talked about religion, family, baseball, movies, books. She was the first smart person I'd ever met who shared my loathing for *Crime and Punishment*—Raskalnikov's endless, infernal pacing back and forth in his little hovel, the maddening profusion of ellipses that were Dostoyevsky's punctuation of choice.

"I wanted to throw the book at the wall," she said.

"I quit reading with eighty pages left," I said.

"Reading it to the end is its own form of punishment," she said.

"Maybe I read it too young," I said.

"Maybe I read it too old," she said.

"Maybe we're a couple of philistines," I said.

"Maybe we need a better translation," she said.

Out of an instinct for self-protection, I resolved to believe she must have a boyfriend. A woman like her couldn't possibly leave her apartment in New York without drawing the attention of every sentient man. This intuition—that her romantic affections were spoken for—relaxed me somewhat, but I still didn't feel quite myself.

Around midnight, her roommate, an aspiring actress with a lilting British accent, pried herself free of the village drunk and got up to leave. I figured Martha would leave with her, and that would be the end of it—but no, the hugs were commencing, the goodbyes and are you okays and I'll see you tomorrows, and then Julia was gone but Martha was not. She took a sip of her drink, turned to me with a serious look on her face, and I thought, Okay, here it comes, the moment when she says she's really enjoyed our conversation, asks if it's all right if we just be friends, and I'll say, *Yes, of course,* and if I'm lucky I'll get to be the indulgent, sweet-natured, unaggressive guy with whom she has coffee every other month or so.

"I have one question," she said.

"By all means."

"Is there going to be any kissing?"

I was stunned; I hesitated for an unconscionably long moment, and by the time I began to stammer in reply, she was so mortified she'd grabbed her purse and excused herself to the ladies' room. She was gone several minutes, long enough for me to move through the proverbial five stages of grief. First came shock and denial—I couldn't have been that clumsy, could I? This was followed by pain and anger—what a moron I was! Then came the shift to bargaining: *If only I can have one more chance, I'll get it right, I'll seize the day . . .* A period of depression ensued, during which I became certain I'd blown my only opportunity; eventually I achieved a measure of acceptance and even a slim measure of hope.

When she returned, she settled her bill, which was not her bill at all but her roommate's, and then she turned to leave with a mumbled goodbye. I'd composed myself sufficiently to utter the words I'd been practicing in her absence. I told her I had an extra ticket to a jazz show but no one to go with me: two nights from now, on Sunday. Dave Douglas at Blue Smoke, potential to be magical. Would she care to join me, maybe have dinner beforehand?

"Sure," she whispered, turned to go.

"Wait," I said. "I don't have your phone number, or any other way to reach you."

She scrawled it on a napkin. I tucked it in my wallet, looked up, and she was gone.

Mercifully, she answered my call the next day, and we spent most of that month together, every spare moment we had—laughing, walking the city, eating, drinking, rarely sleeping, telling stories, laughing some more, listening to jazz. We exchanged exquisite novels with complicated titles. I gave her *So Long, See You Tomorrow* by William Maxwell. She gave me *The Last Report on the Miracles at Little No Horse* by Louise Erdrich. Halfway through that whirlwind month I broke the news of my impending departure to a land without telephones. I expected her to say that was the end of it. I was so gaga-gone I'd even prepared a rebuttal: I would give up my fledgling career as a lookout. I would stay in New York. I would find another job. I would do anything, in fact, to be with her.

"I'll be here when you get back," she said. "We'll write letters. It'll be deliciously Victorian. All the old-school restraint and coded language."

The language wasn't all that coded, but the restraint imposed by a distance of 2,000 miles was quite real. On my days off I hustled down the mountain and into town to pick up my mail. Amid the many letters I often found an overnight FedEx box, aromatic with Irish soda bread or peanut butter cookies, even once a batch of *bizcochitos*, the state cookie of New Mexico. Our correspondence became epic, as much as four letters a week, each of us attempting to amuse the other with picaresque tales—Martha's involving the absurdities of office work and city life, mine of a seasonal job in the United States Forest Service. Hers were entertaining, often touching, and always beautifully written. I took the opportunity, once I got rolling, to unburden myself of seven years of thinking about my brother's death by self-inflicted gunshot. How liberating it was not to have to go through his whole story face-to-face with a potential romantic interest, a story I'd often cut short the moment I detected a look of pity forming on the listener's face. "All families of suicides are alike," the writer Janet Malcolm has claimed. "They wear a kind of permanent letter S on their chests. Their guilt is never assuaged. Their anxiety never lifts. They are freaks among families the way prodigies are freaks among individuals." In my letters I flaunted my S to Martha, practically forced her to trace it with her fingers as she traced my words on the page, and to my good fortune she did not back away in fright.

The following summer Martha quit her job in the corporate office of America's premier provider of theme park entertainment to join me for most of a season in the lookout. That we could survive both a summer of contact only by letter and a summer stuck together in a two-room cabin in the wilderness indicated that we might have been destined for each other, so we returned to New York, packed our belongings into a big Penske truck, and moved to New Mexico for good. The next spring I used the occasion of her first visit of fire season, during her spring break from nursing school, to genuflect on one knee in the tower at sunset and propose she marry me. Her own knees were still raw and bloodied from a scramble through the snow to reach the mountaintop; she likes to say I proposed when I realized there couldn't be many women on earth willing to crawl on all fours to enjoy my company.

Over the years she has gracefully tolerated my desire for a season of solitude, visiting the mountain when time and energy permitted as she reinvented herself as a registered nurse. Aware that my summer work allows me to nap on the job, my winter work invites me to drink on the job, and that neither provides health insurance or retirement savings, she undertook the task of providing for our partnership those perquisites of modern living. If there's a wife more indulgent of husbandly eccentricity, I do not know of her.

Each spring she reminds me I once vowed I'd find a proper line of work that didn't require my disappearance for ten-day stints all summer. I know my long absences aren't easy for her, and we've had our share of pained conversations on the subject. My career, once so novel and romantic, becomes, for her, a little less so with every passing year. While I've found a pleasing symbiosis between my summer job and a summer vacation, she's made known her preference for a slightly more high-class getaway, to Vancouver maybe, or Paris, someplace with museums, nice restaurants, running water. In moments of frustration she's even been known to call my lookout work and backpack trips "little boy games," a charge for which I have no ready defense. Ultimately, though, she offers me her blessing every spring, and I marvel at my good fortune to have found both the job and the woman of my dreams.

In mid-May the forecast finally calls for rain—and with it a chance of lightning. With the coming of storms an almost unseemly expectation of fire arises in my optic nerves. A tingle begins in the backs of my eyeballs and radiates out to the tips of my fingers and toes. I become less a passive watcher of clouds than a partisan—I want massive cumulus buildup, dark-bottomed thunderheads, my radio antenna sizzling in the supercharged air. I want lightning and I want smoke.

The moment of truth fails to materialize. Instead there are showers, mild and muffled thunder, mostly of the cloud-to-cloud variety. A cool drizzle all day and little in the way of lightning. By midafternoon my rain gauge measures a quarter inch. Mist rises up out of canyons south and east, drowning the mountain in fog. The stealthy hush of it would have startled me once, but having studied weather here for going on eight summers I can see it coming now, a swift-moving glacier of water vapor gathering and charging uphill. All of a sudden the world is gone, my service to the U.S. government rendered moot. I abandon my post without remorse.

Martha arrives, her presence announced by the excited yips of the dog. A lovefest commences between them in the meadow, a reunion of licks and kisses that quickly moves indoors due to Martha's need for warm, dry clothes. I stoke the fire and help her unload her pack: some good bread, cheese, chocolate, fresh fruit, a bottle of Zinfandel, and two beers. There is much to celebrate. She's soon to be a nurse: she earned her degree last week and has a job lined up back in town, beginning in July. Our hand-to-mouth existence is about to come to an end. We may even venture an upgrade to the 1988 Dodge Caravan that no longer runs in reverse.

Despite having been alone here for a good long while, I move with ease and even joy into the domestic realm. One of our pastimes involves collaboration on a big pot of chili, and today is the perfect day for it. One of us begins with sautéed onions, a little garlic, some sausage browned in the skillet. We take turns adding this and that over the course of several hours, until we've got it right—or as right as it can get with the spices on hand. While the pot bubbles on its blue flame we play cribbage, notching our respective victories along opposite edges of the board, our very own World Series and almost as intense, minus the 50,000 screaming fans. In cribbage the scoring happens so quickly that keeping track with pen and paper is impractical, unless you're waiting out a snowstorm in a remote mountain cabin, where any old method will do. The horse-race element of the lap around the board with leapfrogging pegs adds a frisson of excitement and encourages trash talk. Our board is a curious piece of work. Made some years ago by Mandijane's husband Sebastian on a similarly fogged-in day, it was fashioned from a length of two-by-four two feet long, the peg holes bored out with a cordless drill we keep on the premises. Each lane is painted a different color—red, white, green—and so are the matching pairs of nails we use for pegs. The board could easily double as a billy club; you could use it to beat a rabid fox to death if you had to.

Amid the drowsy-making heat of the woodstove, our concentrations suffer and we abandon the fourth game midway through. One of us drifts to the bed with a book. One of us fixes a snack. The dog gets a treat and a scratch behind the ears, the humans a guilt-free nap. This is about as civilized as you could hope to live out here in the wild, though the mist will eventually grow tiresome. Past a certain point it can no longer be called mist; it's clear what we are is lost in clouds. For twenty-one hours we are denied a view beyond the meadow on top. The cloud ceiling exists somewhere far below us. Visibility extends a hundred yards. If Martha weren't here I might go a little nutty from the lack of a view, a view being the thing you'd assume I could take for granted up here. Sun and sky and distant desert are my media. When they disappear I'm like a sailor trapped inland, far from his boat. This time, though, alone with Martha on our island in the sky, I'm happy to be marooned.

OUR IDYLL IS OVER in forty-eight hours. Duty calls to Martha in the world below. Her departure always sends me into a funk. Adjusting to her presence here is simple; accepting her sudden absence is a foul and brutal process. I mope about the mountain tossing stones at trees, gathering wood, staring forlornly at the far-off desert. I feel raw and tender in the middle of my being, as if one of my ribs has been wrenched loose with rusty pliers. Only the dog offers solace.

I let Alice out of the cabin to play our favorite game, in which she jumps for a stick I hold aloft at shoulder level. She quickly tires of this and wanders off, sniffing around the meadow. I turn my back on her for less than a minute; when I glance again in her direction, I catch a glimpse of her hindquarters disappearing at a dead sprint into the tree line. I know what this means. She has caught Martha's scent and she's not coming back. I've been careless; this has happened before. After Martha's visits I typically keep Alice locked down in the cabin for at least two hours, until she forgets that Martha was ever here. In my melancholy I abandoned protocol. I know it's too late to call her back or catch her now. She won't stop running until she comes upon Martha, and I won't see her again until I return to town for days off. My funk deepens: I've been repudiated, my company judged inferior. Or maybe Alice simply likes the thrill of the chase. Either way it stings.

That my feelings could be hurt by a dog—that I do not greet her flight as a liberation into purer solitude—is evidence of a revolution in my attitude toward domesticated pets. Until we adopted Alice, I'd never felt much of an impulse toward dog ownership. Experience with the dogs of family and friends indicated they were odoriferous, overbearing beasts, dedicated to immediate gratification of whatever urge bubbled up in their tiny little brains, their owners perversely in need of unconditional love and mindless diversion. But Martha kept telling me her existence felt unnatural without a dog—she'd had one all through her childhood and most of college—and what kind of husband would I be to force an unnatural existence upon my wife, or at least more unnatural than the one I've already foisted on her? My hundred-day sojourn on a mountain each summer makes our marriage unusual enough. I suppose in some way Alice represented a compromise, whereby I'd continue to be that rare creature, a married lookout, and Martha would be compensated with a canine companion in the family unit. Now that Alice has been in our lives for three years, I see her for what she truly is: an odoriferous, overbearing beast dedicated

109

to immediate gratification of whatever urge bubbles up in her tiny little brain, and a reliable and even comforting source of unconditional love and mindless diversion.

Also, she's pretty cute.

FOR WEEKS I'VE BEEN DREADING the arrival of my relief. A rookie named Ben, he will require a thorough tutorial in the art of lookoutry. I've never considered myself much of a teacher, and the skills required of a person here, aside from the use of the Osborne Firefinder, are more intuitive than mechanical and therefore difficult to impart. It's one of those jobs you can learn only by doing. Plus I've become accustomed to having the facilities all to myself, each tool hung on its proper hook; now I'll be forced to share the premises. This place may be part of the public domain, open to visitors any hour of any day, but I've become possessive of it, and anyway the day hikers never stay more than an hour. The arrival of my relief is something altogether different. It's as if I'm about to take vacation and surrender my house to a stranger. Who knows what debauched rituals will be enacted in my absence?

On the other hand, visitors bring the power to lift me from my loneliness, so when my young colleague and his mule packer radio shortly before their arrival, my spirits brighten. I descend the tower and greet them in the meadow, Les again on horseback, leading a single mule, Ben on foot looking a little bit whipped and a good bit lost.

I'd heard that he was a raw youth mere days out of high school—thus his late arrival, nigh on June—but I'd chosen to picture him a budding scholar, introverted but likable, at ease in his own skin and wise beyond his years: basically, the polar opposite of myself at his age. Instead what I find is a gangly kid hardly in need of a proper shave for more than one square inch of his chin, his eyes betraying the pained realization that he now resides far from the comforts he has taken for granted as an American in the twenty-first century. After we unload his supplies and stack them on the porch, his first move is to wander around the meadow with his cell phone in hand, searching for a signal. "It keeps showing two bars," he says, "but it won't let me send a text."

"Now and then a day hiker claims to get a signal, but I think a lot depends on your service provider."

"I thought for sure I'd be able to text," he says.

"It's only four days you're out of reach."

"Yeah, but I've got a girl down in El Paso."

"Invite her to visit."

"She don't hike."

I look at Les. He rubs his beard and shrugs his shoulders, climbs up onto his horse, says goodbye and good luck. He aims to beat feet ahead of another round of storms. In my head I'm already calculating the odds of the kid's lasting the month of June, much less the whole season. Shy of even money, I figure, studying the slightly haunted look in his eyes.

Nonetheless I proceed as if he's here to stay. I can be an ingratiating host—the residual habits of the professional bartender, to which I have access at any moment. Sit right down. Make yourself at home. Let me pour you a drink.

I lead him on a ten-minute tour of the mountaintop—the cabin, the outhouse, the old corral, the likeliest places to collect firewood. I show him where to find all the tools he'll need, how to hook up a new propane tank if one goes empty. I demonstrate the use of the cistern. I point out the stash of extra toilet paper. I help him stow his food in the cupboards. Among his supplies are three cases of bottled water.

"Didn't anyone tell you about the cistern?"

"I guess they mentioned it, but they never said it was safe to drink. Besides, I don't like regular water."

He holds up a single-serving packet of Kool-Aid drink mix, demonstrates for me the magical marriage of sugar-free flavor crystals and mineral-enhanced purified water in a see-through, petrochemical container. The color of it is striking, the concept pure genius: entice a person to pay for something he can find for free by providing it in a package ergonomically designed to fit in his hand and perfectly engineered for absorbing an additional value-added product without the need for a pesky spoon. God bless America, land that I love.

We climb the tower and I begin the spiel I've practiced over the years on the curious and the uninitiated. I point out the major landmarks. I demonstrate the use of the Osborne Firefinder. I explain the principle of triangulation, how you cross your azimuth line with that of another lookout using the strings on the map. I enumerate the features of his radio: the various channels and their purposes, the scan function, the knobs and toggle switches. It's not long before he's plainly bored. He graduated from high school five days ago; he's through with teachers and lessons and for that it's hard to blame him. I must seem another old-timer possessed of a trove of dubious knowledge, enthralled by all the little things I know and he doesn't.

"That hike wore me out," he says. "If you don't mind I think I'll head down and take a nap here in a bit."

"Fine with me."

"Pretty good view, though."

"That's why we're here."

For a while we sit in silence. Along the Black Range storm clouds swell, cumulonimbus rising like misshapen bread dough, their shadows darkening the crest. I can feel their power deep in my bones—the stillness that descends in their gathering a kind of deception, the proverbial calm before the storm. Les radios from the pass, lets us know he rode back safely. The animals are in the trailer and he's headed down the road. I use this little exchange as a teaching tool in the proper use of the radio. Ben seems attentive, if a little overwhelmed.

"Don't worry. We'll go over it a couple more times in the morning. With a little practice you'll be an old pro."

"Yeah," he says.

The first lightning strike hits, the flash and the crash damn near synchronous. I see it to the north of us a quarter to a third of a mile, a livid filament. Ben claims to have seen it to the south of us about the same distance. Either it was a forked strike and we each saw one of its fingers, or it was a single strike and one of us was fooled by its reflection in the windows of the tower. Either way it struck awfully close.

"I think I'll take that nap now," Ben says.

He hotfoots it down the steps like a man late for a court date. I'm not inclined to judge him for that. The first time I saw a strike so near the tower I nearly shat myself, and where there's been one others often follow. I sit on the cot with my knees tucked under my chin, swiveling my head this way and that to follow the progress of the storm. Strikes pound the ridges to the south and east, and each one makes me twitch. I have heard more than once the story of a lookout who was sitting in this tower when it was zapped by lightning, and though the structure is grounded with copper wires running to bedrock, the force of the energy through its metal frame blew his shoes off his feet and knocked him unconscious for something like five minutes. Some who knew him claim that ever after he was not the same man. I have no desire to repeat the experiment—I'm comfortable with my eccentricities as they stand—but I'd hate to miss a smoke the instant it showed. I choose to play the odds and hope. Fingers crossed; sphincter clenched.

The storm moves off. The clouds break apart. An hour passes without incident. I look at my watch and see I've stayed past quitting time. I hang my binoculars on their hook, tidy my stack of radio logs, on which I note each communication to and from my post, and prepare to knock off. Then I see it. Nearly due north, six hundred yards at most. A single lightning-struck tree sending faint tendrils of blue smoke up the ridge. I consider rousting Ben from the cabin for a real-time, hands-on lesson but decide against it. The smoke will still be here when he wakes, and I'll guide him through the process then.

I plot the fire's location—very simple from this distance—and call it in to dispatch. It smokes at the highest reaches of the Cobre Creek drainage, so I name it the Cobre Fire. Because of the waning daylight and the fire's considerable distance from a road, my superiors decide they'll attack it in the morning with smokejumpers, winds permitting. With the forest damp from recent rains, the fire is unlikely to spread. A cool night of high humidity will deny it the conditions it needs to thrive. Come morning it will hardly be alive.

SMOKEJUMPERS HAVE A LONG HISTORY on the Gila. Since 1947, a contingent has been stationed just south of the forest, first at Deming, then later at the aerial fire base near Silver City. They arrive each spring from bases up north—Missoula, Montana; McCall, Idaho; Fairbanks, Alaska—and spend a couple of months here. The elite corps of wildland firefighters, they're dispatched on initial attack to smokes in some of the most remote and difficult country in the West, places beyond the reach of roads, places to which it would take a day or more to get a regular crew on the ground.

The first jump of a wildfire occurred in July of 1940, in the Nez Perce National Forest in Idaho. The men tossed burlap sacks out of the plane's cargo hold in order to test the winds; they jumped wearing padded leather football helmets with wire-mesh face masks to protect against head injuries. Many of the earliest smokejumpers were conscientious objectors during World War II. Opposed to killing men in Europe and Asia but not to serving their country, sixty of them happily signed on to fight fires in the forests of the American West. Their experiments were closely monitored by the U.S. Army, which borrowed the lessons learned in jumping wildfires and applied them to the formation of paratrooper units, such as the 101st Airborne.

In addition to possessing a colorful history, the smokejumpers were blessed to have a poet write a beautiful book about them. Norman Maclean's *Young Men and Fire* is the one and only masterpiece ever written on the subject of American wildfire. Better known for his novella *A River Runs Through It*, which Robert Redford made into a movie starring Brad Pitt, Maclean worked summers with the Forest Service when he was only a teenager, an opportunity he was afforded because able-bodied woodsmen had gone to fight in World War I. In 1920, Maclean went east to college at Dartmouth, to study English. When he finished, he returned home to Montana and worked again briefly in the mountains of his youth. It was a moment that divided his past and his future forever. He often looked back at a career that might have been—a career in the woods, in logging camps, fighting fires and packing with mules and playing cribbage in the bunkhouse. Instead he went to the University of Chicago, where he earned his doctorate and later held the post of William Rainey Harper Professor of English. He taught there for more than forty years, mostly Shakespeare and the Romantic poets, and each summer decamped from Hyde Park for Montana, where he spent three months at his family's cabin on Seeley Lake—not far from where the smokejumpers made some of their first practice jumps in the summer of 1940.

Nine years later, Maclean was home again for the summer when he heard about a fatal wildfire in Mann Gulch, a steep side canyon near the breaks of the Missouri River. Thirteen smokejumpers had died when a fire they were fighting blew up below them. A few days later Maclean arrived at the scene of the conflagration, with the fire still burning in stump holes and scattered trees. The smell and the look of the gulch haunted him the rest of his life. After his retirement in 1974, he would spend years attempting to reconstruct what had happened to the men who died there. *Young Men and Fire* is, in one sense, the story of an unforeseen disaster, in which smokejumpers accustomed to unifying earth, wind, and fire found themselves overwhelmed by that final element, which they could not outrun. It is also, as Maclean put it, a story in search of itself as a story—or, to say it another way, a tragedy in search of a tragedian. Late in the book he writes:

> Those who know something about the woods or about nature should soon have perceived an alarming gap between the almost sole purpose, clear but narrow, of the early Smoke-jumpers and the reality they were sure to confront, reality almost anywhere having inherent in it the principle that little things suddenly and literally can become big as hell, the ordinary can suddenly become monstrous, and the upgulch breeze can suddenly turn to murder. Since this principle comes about as close to being universal as a principle can, you might have thought that someone in the early history and training of the Smokejumpers would have realized that

something like the Mann Gulch fire would happen before long. But no one seems to have sensed this first principle because of a second principle inherent in the nature of man—namely, that generally a first principle can't be seen until after it has been written up as a tragedy and become a second principle.

To tell the story, Maclean taught himself the latest wildfire science with dogged precision, coaxed a friend to help him mark time on the hill, and read and reread the official report on the fire. He plumbed the recollections of the fire's three survivors, two of whom he led back to the scene to revisit their close encounter with death. The smokejumpers' foreman, Wag Dodge, had lit an escape fire and lain down in its ashes as the big fire whirl passed over his men: he tried in vain to persuade them to join him, the only hope for survival most of them had, though none of them listened. "With the flames of the fire front solid and a hundred yards deep he had to invent the notion that he could burn a hole in the fire," Maclean wrote. "Perhaps all he could patent about his invention was the courage to lie down in his fire. Like a lot of inventions, it could be crazy and consume the inventor. His invention, taking as much guts as logic, suffered the immediate fate of many other inventions—it was thought to be crazy by those who first saw it."

Dodge would die five years later, still haunted by his inability to show his men the way to their afterlife through and beyond the fire. Maclean returns repeatedly to a version of the event as a kind of Passion play, with Stations of the Cross scattered up the hill, marked now by literal crosses where each of the dead men fell, monuments to their unimaginable end. Although Maclean never says so explicitly, Dodge resembles a kind of Jesus figure, misunderstood in his message at the moment it counted most for the world, that world for Dodge consisting entirely of young men running uphill in a gulch and a wildfire running faster forever and ever—the nightmare scenario of all who have ever fought fire.

Though he worked on the story for most of the last fourteen years he lived, the manuscript remained unfinished upon Maclean's death. When it was published in 1992, *Young Men and Fire* was awarded the National Book Critics Circle Award—a sadly posthumous tribute to a genius of American letters whose only other book was short-listed for a Pulitzer in 1977, a year when the judges deemed no book worthy of the award and abstained from offering a prize in fiction. But having started writing so late in life—he began *A River Runs Through It* only after he retired from teaching at the age of seventy—he was never angling for prizes. He was angling for immortality, and with both of his books he achieved it.

THE NEXT MORNING THE COBRE Fire hardly shows any smoke. Nevertheless, the jumper plane lifts off from Grant County Airport at 8 a.m. It circles the fire for half an hour, the spotter assessing likely jump spots and dropping paper streamers, testing the wind. The streamers unfurl and float into the treetops, bright pennants of yellow and pink, weighted slightly on one end with a pouch of sand to ensure they don't just float for miles. Their colors make them easier to follow than those original burlap sacks, and the weight of the sand and the length of the streamer are calibrated to mimic the drift of the average-size smokejumper. The spotter judges the winds light enough to jump. The open meadow on Apache Peak provides the target for landing. From our privileged perch, Ben and I watch as first one jumper and then another leaps into midair, chutes abloom above them in an instant. They swoop toward the mountain amid swirling ridge-top winds, soaring like giant birds.

One of them, fooled by a sudden wind gust, slams into the meadow with an audible thud and performs an awkward half-somersault. The other lands smoothly but manages to hang his chute in a tree. Their para-cargo—food, water, tools, tents, sleeping bags—follows soon after in a box with its own chute. I amble over to greet them, make sure they're still in one piece. The first of them beams at me, still geeked up on the thrill of the leap. "Dawson," he says, extending a hand. "Not very often we jump at this elevation. Those ridge-top winds were gnar-gnar"—jumper shorthand for gnarly times two. I tell Dawson what I know of the fire and how to get there, while Ben helps the other jumper, Chris, untangle his chute from the branches of a wind-stunted pine. We learn they're BLM jumpers from Alaska. In half an hour their gear stands neatly stacked in the meadow, and they're ready to begin their hike to the fire.

"Looked from above like we could take a leak on it and that would be that," Dawson says.

"Cocktails and cribbage at six, then?"

Chris smiles and flips me the thumbs-up.

"Deal me in," he says.

All day long I scope the surrounding country for sleeper smokes, snags hit by lightning that may have smoldered overnight. None appears. By early afternoon the Cobre Fire no longer shows. I try to engage Ben in conversation, but it becomes clearer by the moment that, being half my age, he regards me as something of a bewhiskered fogey. That I would write letters to friends on a typewriter, sit with them for days making notes in the margins, hike down the mountain with them and drop them in a mailbox to be trucked overland across the country, strikes him as about as antiquarian as sending word by smoke signal or semaphore. He becomes animated only when I ask about his high school's rivalries with other local towns. He tells me he and his buddies in Truth or Consequences, better known in these parts as "T or C," or simply Torc (rhymes with *dork*), have a particular hatred for the boys of Hatch, a town just down the road that proudly calls itself the green-chile capital of the world. After football games the boys of each town honor an unspoken agreement with their rivals to gather in the parking lot of the Sonic drive-in, dump Jim Beam into their Styrofoam soda cups, say unkind things about the size of each other's genitalia, and, when properly lubricated, "throw down on the little sissy bitches," as Ben puts it. Good, clean American fun.

Around four o'clock, Chris and Dawson tramp into the meadow. Their fireproof Nomex uniforms look as fresh as the day they were laundered. They laugh about how easy a fire it was to put out. "A few shovelfuls of dirt and it was nap time," Chris says. Hardly the stuff of tragedy or even poetry—more like farce, that the Forest Service, in a spasm of institutional habit, would drop men from an airplane on a fire that would have burned out on its own after skunking around in a stump hole for a day or two.

Yet I have been provided with a skilled cribbage player, and for that I'm grateful. Dawson and Ben decline my invitation to the cocktail hour, but Chris grabs the deck and begins to shuffle. I simmer a packet of creamy noodles to supplement Chris's Top Ramen. We find a ball game on the radio, hoist a glass of spirits, toast our benefactor: *To the United States Forest Circus, aviation and prevention divisions.*

"Dude, the jumpers back at the base are gonna hate me," Chris says. "Whisky, cards, chocolate, a baseball game—this is the cushiest jump I've ever made."

"Don't forget coffee in the morning," I remind him. "Freshly ground, organic, grown in the shade."

"We should've called the fire a half-acre and stayed another night."

FOR FOUR DAYS I LEAVE the mountain to Ben. Even as I reacquaint myself with the pleasures of "syphilization," as Edward Abbey called it, a part of me can't help but wonder how the kid is holding up. Storms roll through all weekend, and the local newspaper, the *Silver City Daily Press*—or *Daily Press Release*, as it's more accurately known—reports several new fires on the forest. I'm tempted to leave my radio on all weekend, scanning for news. Because we live on a hill south of downtown, I can make out the dispatcher, Cherry Mountain, and Apache Peak too if I set my radio on a bookshelf in the living room. For a day and a half I resist the temptation but on Saturday afternoon I break down and eavesdrop for a few minutes. From what I can gather in bits and pieces of half-broken transmissions, it seems a fire crew is lost in the Black Range, somewhere north of Apache Peak. They've run out of water, so a helicopter has been dispatched to drop them some and save them from dehydration. Ben sounds panicky and overwhelmed, the dispatcher exasperated, the helicopter spotter bemused. This is what my Forest Service friends call a Class C clusterfuck. I can't bear to listen very long. Later I will learn that the crew thought they could hike cross-country from their fire to Wright's Saddle, a distance of seven miles across tough country, traversable only by mule deer and elk.

I mention none of this when I return to the lookout on Tuesday. Ben is sitting on the porch when I arrive, his bedroll looped to his pack. Our transition, totaling maybe three minutes, is curt and sort of sad.

"I'm still young," he says. "I need to be around people. I got so lonely I hiked out Sunday night and drove into town. I hiked back up the next morning."

"Are you cashing it in for the summer?"

"I'm thinking about it."

Thinking about, hell. His pack is enormous. He's cleaned all his canned food out of the pantry, leaving only a few packets of dried fruit and the bottled water. He's burned more wood than he cut, committing one of the cardinal sins of lookoutry. I can't work up much in the way of outrage, though. Almost every year one or two souls spend a day in a tower on the Gila and never return. Just last year one guy got on the radio his second night on Loco Mountain, not long after dark, and called the Beaverhead work station to say his heart wouldn't stop pounding. It took half an hour of coaxing over the radio by the station manager to calm him down, and the next day he hiked out and was never heard from again. When you consider a person has to be free of a fear of fire (*pyrophobia*), a fear of confined spaces (*claustrophobia*), a fear of being alone (*isolophobia*), a fear of heights (*acrophobia*), a fear of steep slopes and stairs (*bathmophobia*), a fear of being forgotten or ignored (*athazagoraphobia*), a fear of the dark (*nyctophobia*), a fear of wild animals (*agrizoophobia*), a fear of birds (*ornithophobia*), a fear of thunder and lightning (*brontophobia*), a fear of forests (*hylophobia*), a fear of wind (*anemophobia*), a fear of clouds (*nephophobia*), a fear of fog (*homichlophobia*), a fear of rain (*ombrophobia*), a fear of stars (*siderophobia*), and a fear of the moon (*selenophobia*), then it's little wonder most people aren't meant to be lookouts.

As Ben lumbers off under the weight of a giant pack, I know I've seen him for the last time. Unless my superiors find another relief lookout in a hurry—unlikely, given the prerogatives of government paperwork—more extended tours and overtime await me. I can't say I'm disappointed by this. Martha's about to go east and visit family for a month, an annual ritual, and I'll be left with little for which to pine back in town. Instead I'll reapply myself to the unfinished project of learning how to think like a mountain. And if the silence begins to disquiet me, I can listen in on my radio's tactical channels. The west side of the Black Range has caught fire in my absence—two new lightning starts called the Diamond and the Meason—and for the next several weeks their smoke will hug the mountains to my north. These fires, it's been decided, will not be suppressed. The time has come for the forest to burn.

3

JUNE

Man always kills the thing he loves, and so we the pioneers have killed our wilderness. Some say we had to. Be that as it may, I am glad I shall never be young without wild country to be young in. Of what avail are forty freedoms without a blank spot on the map?

—Aldo Leopold, *A Sand County Almanac*

Aldo Leopold & the world's first wilderness
✳ the joys of walking in cities & moun-
tains ✳ a look back at the McKnight Fire ✳
Meason & Diamond fires on the move
✳ fish rescue in South Diamond Creek ✳
camping in Apache country ✳ scenes from the
Victorio War ✳ the wisdom of Marlin Perkins

MONG THE HIGH HOLY DAYS I observe each summer, the third of June holds a special place. On this day in 1924, Aldo Leopold's plan to preserve a roadless core surrounding the headwaters of the Gila River received approval from the regional office in Albuquerque. The proposal had been under consideration for two years, during which his draft memo and the accompanying maps were for a time misfiled and presumed lost. From a logistical perspective alone, the ruling was impressive: it banned roads, auto travel, hotels, summer homes, and hunting lodges from an area nearly the size of Rhode Island. From a cultural standpoint, though, Leopold's vision for the Gila marked a decisive turn in America's treatment of wilderness. Never in history had a government body, American or otherwise, surveyed the values of so large a piece of country and decided that its highest use lay not in economic exploitation or—as in the case of the national parks—scenic wonders, but in no use at all except by the nonmotorized traveler. If country can be thought of as a text, then the Gila ought to be considered the first rough draft of the wilderness prospect in America.

Two forces shaped Leopold's thinking as he mulled his proposal for the Gila Wilderness—one practical, one intellectual. His inspections of the national forests in New Mexico and Arizona offered the practical perspective. When he'd first seen the territories in 1909, six areas of between half a million and a million acres contained no roads. In Arizona, these included the Tonto Basin, the Kaibab Plateau around the Grand Canyon, and the Blue Range and adjacent White Mountains. In New Mexico the same held true for the Jemez and Pecos divisions of the Santa Fe National Forest and the Gila River drainage of the Gila National Forest. By 1919, roads had punctured all of them but the Gila. Only the rough and broken nature of the country had spared it.

The second major influence on Leopold's thinking occurred when he met a young Forest Service architect in December of 1919. Arthur Carhart, like Leopold a native Iowan, was the first full-time landscape architect to be employed by the agency. Among his early assignments was to survey the shoreline of Trappers Lake in the White River National Forest of Colorado, a site of sublime alpine beauty, in preparation for the development of summer homes. Carhart returned from his mission with a surprising proposal: leave the lake alone. Summer homes would only mar it.

A colleague of Carhart's, aware of the preservationist instincts he shared with Leopold, arranged for the two of them to meet.

Their discussion had an impact out of proportion to its footnoted place in the history of American conservation. Though they were hardly the first to express concern over the disappearance of wilderness, their meeting of the minds galvanized an effort within the Forest Service to preserve wilderness before it vanished. While Leopold turned his attention to what remained of the Southwestern wilderness, Carhart became a proponent of protecting what we now call the Boundary Waters Canoe Area Wilderness of northern Minnesota.

Carhart later recorded the essentials of their talk in a memo for Leopold: "The problem spoken of in [our] conversation was, how far shall the Forest Service carry or allow to be carried man-made improvements in scenic territories, and whether there is not a definite point where all such developments, with the exception perhaps of lines of travel and necessary signboards, shall stop." Their shared thinking centered on the idea of limits: "There is a limit to the number of lands of shore line on the lakes; there is a limit to the number of lakes in existence; there is a limit to the mountainous areas of the world, and in each one of these situations there are portions of natural scenic beauty which are God-made, and the beauties of which of a right should be the property of all people."

Starry-eyed notions of wilderness conservation far predated Leopold's vision for the Gila. In 1810, the poet William Wordsworth described England's Lake District as a "sort of national property in which every man has a right and interest who has an eye to perceive and a heart to enjoy." In 1832, the painter George Catlin, famous for his richly colored portraits of Plains Indians, argued in favor of preserving a "nation's Park" on the northern prairies, a kind of museum of the natural world, to include both wild animals and native people. Forty years would pass before the creation of anything resembling Catlin's idea. By the time the U.S. Congress voted to establish Yellowstone National Park, indigenous people almost everywhere had been killed, imprisoned, or shunted onto marginal reservation land. Yellowstone was saved mainly in the interest of preventing private exploitation of its geothermal wonders. Those who fought to protect it thought of it as a collection of natural curiosities, not a functioning wilderness. The railroads, after coveting the park's resources for years, saw a chance to profit nonetheless: if they couldn't be given access to the timber, they could at least transport, feed, and shelter visiting tourists from the East. Camping out of doors for the pleasure of it remained an eccentricity, and the park was duly developed with roads and hotels to serve the masses. New York's Adirondack Forest Preserve, established in 1885, also found its reason for existence in a utilitarian rationale: it offered protection to the major water source of New York City.

As Leopold developed his plan for protecting the Gila, he tried to synthesize two divergent schools of conservationist thought. One of them, exemplified by the bearded sage John Muir, advocated preserving wilderness both for its own sake and as a natural cathedral for the human spirit, where man could come face-to-face with higher values. "Mountain parks and reservations are useful not only as fountains of timber and irrigating rivers," Muir wrote, "but as fountains of life." He called the forests of California "God's first temples," and spoke in rapturous prose about trees "proclaiming the gospel of beauty like apostles." When he'd encountered a rare orchid, *Calypso borealis*, on his rambles through the swamps of the Great Lakes region, he'd sat down next to it and wept at its beauty and fragility. He sensed a profound interconnectedness in all of nature. "When we try to pick out anything by itself," he wrote, "we find it hitched to everything else in the universe."

The other school of thought was led by Gifford Pinchot, President Teddy Roosevelt's right-hand man on conservation matters. Pinchot argued that the scenic or spiritual qualities of a landscape were not sufficient reason to protect them. He fought against "a fundamental misconception that conservation means nothing but the husbanding of resources for future generations. There could be no more serious mistake." When scenic values came into conflict with the material needs of the American people, scenery had to give way to "highest use." How far development should go, or who was to decide what constituted the highest use of any particular natural resource—these were slippery questions Pinchot left unanswered.

The two men's differences are neatly illustrated by an event that occurred while they were on a group outing in the Grand Canyon during the summer of 1896. They were part of a commission assessing the forests of the West—some of them already part of the reserve system, others soon to be—and during one of their walks they came upon a tarantula. Pinchot raised his boot to squash it, but Muir stopped him. As Pinchot later wrote in his memoirs, "He wouldn't let me kill it. He said it had as much right there as we did." For Muir, all of creation was sacred, including the gnarliest arachnid; he celebrated "the essential oneness of all living beings." Pinchot believed man ought to be the arbiter of what in the natural world was useful and what was not. Although Pinchot enjoyed Muir's gifts as a storyteller, and they found common cause in their desire to protect the Western forests from abusive commercial interests, they would clash repeatedly in years to come over how those forests should be managed.

Leopold had come out of the Pinchot school of thought—quite literally. He had earned his forestry degree at Yale in 1909. The curriculum there was the first of its kind in the nation, the forestry school having been founded in 1900 with an endowment from the Pinchot family. Oldest son and scion of a timber magnate, Pinchot used the bequest to put his stamp on a generation of forest rangers. The Forest Service in its early years was often referred to as the Washington, D.C., chapter of the Yale alumni association; Pinchot's first rangers, Leopold among them, were called "Little G.P.s" for their loyalty to the chief, not to mention the education that formed them in his mold. The Pinchot doctrine of highest use was summed up in his famous dictum: "the greatest good for the greatest number in the long run." It's worth remembering that Teddy Roosevelt, advised by Pinchot, chose to fold the forest reserves into the Department of Agriculture, not the Department of the Interior, when he founded the Forest Service—a tacit admission that trees and the watershed they maintained were crops to be tended and protected, not unlike corn or wheat.

Leopold's first major job as a professional forester, in the summer of 1909, had been to run a reconnaissance mission on Arizona's Blue Range. The forest reserve there had been established in 1907, and the first thing the Forest Service wanted was a thorough inventory of its new lands. Leopold arrived in time to play his role as one of the "arranger rangers," the group of young men who mapped and cataloged the nation's store of forest wealth and reported their findings to the home office on F Street, back in the nation's capital. Leopold and his crew assessed the type, amount, location, and quality of the timber while mapping and surveying land that largely remained, until their work was finished, a *tabula rasa* to the American government.

In letters home to his parents around this time, Leopold wrote about his joy in the job, about not having to "fight society and all the forty 'leven kinds of tommyrot that includes. [It] deals with big things. Millions of acres, billions of feet of timber, all vast amounts of capital—why it's fun to twiddle them around in your fingers, especially when you consider your very modest amount of experience. And when you get a job to do, it's yours, nobody to help, nobody to interfere, no precedents to follow." The big job that loomed on the Blue, once he completed his survey, involved supplying the copper mines in nearby Clifton with 15 million board feet of timber a year for fuel and mine shaft supports. "I want to handle these 15 million a year sales when they come," he wrote. "That would *be* something."

In the years to follow, Leopold supported efforts to hunt wolves to extinction, believing this would increase the number of deer available to hunters. He favored fire suppression as a means of increasing the timber yield. He advocated draining the Rio Grande Valley around Albuquerque to lay claim to marginal farm land. He argued for stocking non-native trout in Southwestern streams to improve recreational fishing. He personally oversaw some road building projects in the Southwestern wilderness, as part of his agency's "good roads movement." All of this fit with Pinchot's philosophy of scientific management and highest use.

By 1921, the achievements of that philosophy were everywhere evident: roads blasted through the mountains, cattle running the ranges, irrigation works funneling water to farms, predator populations reduced to near nothing. As Leopold surveyed the Southwestern forests, a new and troubling question gnawed at him. Could there be such a thing as too much progress? And if so, what of value would be lost in the bargain?

He first attempted to answer these questions in a 1921 paper, "Wilderness and Its Place in Forest Recreation Policy," a blandly titled essay the effects of which would ripple outward for decades and eventually lead to the creation of tens of millions of acres of protected wilderness in America. Leopold admitted his bias in favor of developing natural resources and even quoted Pinchot approvingly in regard to the doctrine of highest use. "The majority undoubtedly want all the automobile roads, summer homes, graded trails, and other modern conveniences that we can give them," Leopold wrote. "It is already decided, and wisely, that they shall have these things as rapidly as brains and money can provide them." But then he turned the whole concept of highest use on its head. What of that "very substantial minority" of people who wanted precisely the opposite experience of wilderness? Shouldn't the Forest Service attempt to provide for their needs too? Leopold not only answered in the affirmative, he stated that "highest use demands it." Because so little of the original Southwestern wilderness remained without roads, it was imperative that a portion of what was left be preserved in that state, in order to provide a recreational experience that would vanish otherwise.

As an example of the "substantial minority," Leopold pointed to the wilderness packer. "The man who wants a wilderness trip wants not only scenery, hunting, fishing, isolation, etc.—all of which can often be found within a mile of a paved auto highway—but also the horses, packing, riding, daily movement and variety found only in a trip through a big stretch of wild country." The national parks were unlikely to provide such an experience. For one thing, they banned hunting. In addition, "the Parks are being networked with roads and trails as rapidly as possible. This is right and proper. The Parks merely prove again that the recreational needs and desires of the public vary through a wide range of individual tastes, all of which should be met in due proportion to the number of individuals in each class. There is only one question involved—highest use." To conclude his argument, he offered an example of how and where these needs could be met—by preventing new roads in the core of the Gila:

This is an area of nearly half a million acres, topographically isolated by mountain ranges and box canyons. It has not yet been penetrated by railroads and only to a very limited extent by roads. . . . The entire region is the natural habitat of deer, elk, turkey, grouse, and trout. If preserved in its semi-virgin state, it could absorb a hundred pack trains each year without overcrowding. It is the last typical wilderness in the south-western mountains. Highest use demands its preservation.

Thoreau and Muir, among others, had voiced support for wild things and wild nature, and Muir had lobbied tirelessly for the protection of the high Sierras and other scenic lands. But he was, in the words of historian Stephen Fox, a "radical amateur," and lacked the power to impose his vision in any direct way. His method was to plead and cajole and write for a mass audience, hoping to reach the like-minded and stir them to action. Aware of the need for converts to his cause, he had been willing to support roads in Yosemite as a means of bringing the average tourist into contact with the wild. Leopold, on the other hand, worked close to the levers of power, inside a public-lands agency. He had only to convert a handful of his superiors to his vision for the Gila. His proposal arose out of the awareness that "it will be much easier to keep wilderness areas than to create them." In fact, he noted, "the latter alternative may be dismissed as impossible." The ironies and tensions in his writing from this time are palpable: Leopold's concept of wilderness could have been invented only from inside a culture bent on destroying it.

His article, published in the *Journal of Forestry*, received a warm reception. This was partly the result of crude bureaucratic concerns. Many in the Forest Service saw Leopold's view as a means to gain the upper hand in a turf war with the Department of the Interior, which had shown a willingness to usurp scenic Forest Service lands and bring them under its own umbrella as national parks. This had happened with the Grand Canyon in 1919. Others genuinely shared Leopold's concern over the disappearance of roadless country. But some of his colleagues thought his plan was madness; his old boss on the Apache, John D. Guthrie, wrote him a letter of protest: "We are too much getting away from the real forestry idea in this country, and more and more making the national forests into half-baked national parks."

In October 1922, Leopold submitted a formal memo laying out his plan to preserve the Gila headwaters, complete with a set of maps that delineated the wilderness boundary. After dithering for nearly two years, the Forest Service implemented Leopold's plan, and the world's first wilderness was born: 755,000 roadless acres in the heart of the Gila.

Leopold's thinking on wilderness would continue to evolve for decades to come, long after he left the Southwest. He began from a simple scarcity theory of value. To this he added a cultural element, whereby wilderness would allow for the perpetuation of certain skills associated with our vanishing pioneer past, most notably travel by horse or canoe. Eventually he evolved a biological rationale: wilderness should serve as a laboratory of land health, "a base datum of normality, a picture of how healthy land maintains itself as an organism." By the last decade of his life, he had abandoned the Pinchot camp and joined Muir in his belief that wild nature had a right to existence entirely separate from all human claims, whether economic, cultural, or scientific. In *A Sand County Almanac* Leopold wrote:

> It is a century now since Darwin gave us the first glimpse of the origin of species. We know now what was unknown to

all the preceding caravan of generations: that men are only fellow-voyagers with other creatures in the odyssey of evolution. This new knowledge should have given us, by this time, a sense of kinship with fellow-creatures; a wish to live and let live; a sense of wonder over the magnitude and duration of the biotic enterprise.

Above all we should, in the century since Darwin, have come to know that man, while now captain of the adventuring ship, is hardly the sole object of its quest, and that his prior assumptions to this effect arose from the simple necessity of whistling in the dark.

It would be naïve to argue that the Gila Wilderness exists in anything resembling a "pristine" state, whatever that might mean at this stage in history. Parts of it have been chipped away by road-building, despite the original proclamation. A third of it was lopped off by the North Star Road in the 1930s, a road that severed the Black Range from the rest of the wilderness to the west. Several years later another road was built to the Gila Cliff Dwellings National Monument—a picturesque grouping of thirteenth-century Native American homes on the West Fork of the Gila River—in order to grease the tourism trade. Cattle grazing continued in the wilderness for decades, denuding the watershed. Non-native fish were let loose in the streams, decimating natives. Fire suppression altered the cycle of burn and regrowth. The detritus of a human presence litters the landscape: ruined cabins, barbed wire fences, initials carved in aspen trees, potsherds, pictographs. You could even say the presence of my tower, my propane stash, my outhouse, and my cabin diminishes the wildness of the place. You'd get no argument from me.

I've read the literature of the wilderness deconstructionists, so I can list by rote their complaints against wilderness preservation. By encircling wilderness on the map, we actually extended our dominion over it—subdued it for the last time, as it were, by barring it from settlement and dictating the terms of its use or non-use. We imposed an artificial distinction between the human and the nonhuman worlds. We privileged scenic landscapes (high mountains) over more humble ones (marshes, prairies). And we protected places where, in the language of the Wilderness Act, "the earth and its community of life are untrammeled by man," despite the fact that indigenous people lived on those landscapes for millennia until they were removed at gunpoint.

The creation of "managed wilderness" is one of our culture's fundamental paradoxes, and defining why we value it, and how we ought to relate to it, will remain an unfinished project. But leaving aside the notion of simply gifting what's left of it—less than 3 percent of the landmass of the Lower Forty-eight states—to the gas, oil, timber, mining, livestock, and hydropower industries, which at least has the value of a certain stark simplicity, these criticisms never lead to a better idea than the one hammered out by Leopold: that certain samples of our natural heritage should remain, in a gesture of humility to future generations and the nonhuman world, beyond the reach of the bulldozer and the backhoe. "The environment of the American pioneers had value of its own," Leopold would write, "and was not merely a punishment which they endured in order that we might ride in motors."

To mark the birthday of the Gila Wilderness, Alice and I set out on a walk at quitting time, with provisions for an overnight camp: water, matches, a sleeping bag. Seven miles away there's a nice flat spot on a ridge above two canyons, one draining to the north and the other to the south. The trail along the way is almost vanished from lack of upkeep. The notes in the visitor log at the lookout show no record of anyone having been this way in ten years. It's not merely the prospect of guaranteed solitude that attracts me to this wild, windswept ridge. In late May 1922, Leopold broke away from his inspection tour of the Gila to check on a fire that had burned for days, over hundreds of acres, on this very spot. Leopold's colleague Fred Winn, then supervisor of the forest, wrote of the fire in his diary:

5/20/22
Fire is in almost impassable country—awfully dry and rough and water very scarce. . . . My idea is to let the east side burn as it will stop at Cave Creek for lack of fuel and there is less timber there and of little value. . . . The danger is in the south and west on top of Black Range. . . . if fire once gets into [that country], 500 men cannot hold it.

5/21/22
Up at 4 A.M. Painter, Reid and I discussed the situation. They feel that the fire is "corralled" but not safe and rushed a bunch of men out at 6:30 A.M. to take up patrol where the night shift left off. . . . If we can hold the line today, and God Almighty will not send a gale but a westerly wind, we will have it. . . . Leopold came in at 2:30 P.M. Discussed situation with him in detail.

5/22/22
Fire is practically under control according to Painter and Reed. . . . Leopold and Painter left at 3:30 P.M. to ride around fire line. . . . Turned in at fire camp at 9:00 P.M. after Painter and Leopold came back from fire line and figured it safe.

· · ·

FOR THE NEXT MONTH, OFTEN in the company of Winn, Leopold would ride the forest, inspecting the range, assessing the safety of lookout towers, recording the condition of ranger stations and firefighting tools. In what spare moments he had, he looked over the country with an eye toward marking his proposed wilderness boundary. Sometimes, in the evenings, he fished for trout. Then another round of fires broke out—over the course of one ten-day stretch, forty-one new smokes popped up on the forest—and he pitched in to fight them where needed. It was the busiest fire season since 1904. When he returned to Silver City on June 21, Leopold praised the men he'd worked with in the field: "I have seen fire fighters of many kinds in many places, but I believe the settlers and range men of the Gila Forest, with good equipment and supervision, are the most efficient men on a fire line that I have ever seen."

Some of those men, with Leopold at their side, almost certainly trod the ground where I lie in my sleeping bag eighty-seven years later. I try to imagine them in their cowboy hats and bandannas, their faces smudged with charcoal and reddened by the heat of the flames, the banter they shared as they cooked their provisions over the fire at night. Try as I do to hear those voices and see those faces, I come up mostly empty. I turn instead to the earth around me. Claret cup cacti bloom bloodred in the rocks. Wallflowers and penstemon stipple the forest floor yellow, orange, and scarlet. I listen to the wind rustle the shaggy needles of the pines. Alice wanders off in search of small game, heeding her impulse to hunt. Words drift away, images dissolve. I open myself to night sounds and starlight, the sigh and sway of tree limbs, constellations I can and cannot name. A wild and solitary peace, a dream-filled sleep. By dawn the dreams have vanished and my fire is cold. A seven-mile hump stands between my resting place and my workplace. I cannot imagine a sweeter commute.

. . .

A SPELL COMES OVER the mountain in early June. The wind dies, bees hover outside the open tower windows, the cinquefoil blooms butter yellow in the meadow. Clark's nutcrackers flutter their black and white wings as they move from tree to tree. Lenticular clouds dot the highest peaks, their elliptical shapes and striated edges bringing news of howling winds a few thousand feet above me. In the land below them a stillness reigns, and in my little bird's nest I gawk for hours on end. I call in a fire along the Rio Grande, forty miles away, and am told an hour later by dispatch it's a farmer burning his field. The landscape moves through moods all day, morning haze in the valleys, every hollow and river course darkened in shade and swaddled by smoke. Come afternoon, the topographic variations flatten in the absence of shadow. The scene outside my windows leaches of contrast. A searing sun scalds the desert. The cloudless sky seems to open like the lid of a vault, blue to infinity. The job becomes an exercise in just how much watching I can stand. I've yet to find the limit.

This tolerance for watching goes a long way back. As a child I liked to climb the silo on our farm by day and look from the second-story window of our house after sunset. With the onset of dark I could see the solitary yard light that marked the horizon one mile down the road. Over time I discovered I could ignore it, or pretend it was something else, which freed me to imagine all sorts of worlds other than the one where I lived. Maybe the light was a lantern in a lighthouse, and beyond it lay a craggy coastline. Maybe it was a cowboy's campfire low on the flank of a mountain. Maybe it was an oil lamp, hung from a pole where a raft was beached for the night on a sandbar. My secretive nighttime reading of Louis L'Amour and Mark Twain fed these visions, but by daylight a world of geometric precision reasserted itself, row upon

row of corn and beans in perpendicular lines, county roads crossing every mile, the oppressive grid of monoculture farming. All
those replicating corners made me feel hemmed in, claustrophobic—every potential escape route a ruse. As far as I could ride my
bike the pattern repeated ad nauseam, well beyond the horizon
delimited out my bedroom window.

Living in New York years later did much to soften my revulsion at cross-hatched landscapes. Few American places are so
defined by a latticework pattern, and I made it my mission and
found it my joy to trace as many lines of the circuit as I could.
Naturally I found myself attracted to those places where the grid
tilted weirdly or dissolved altogether, the derelict waterfronts for
instance, but even where the grid was at its most rigid I found
pockets of the human carnival so vivid they kept my vertigo at
bay. I can't count how many times I walked across the 59th Street
Bridge and down the island of Manhattan, through the Lower
East Side and around City Hall to the Financial District, ghostly
cold in its nighttime emptiness, and then up through Battery Park
and Tribeca and over to Chinatown and on to SoHo and the West
Village and Chelsea, weaving amid the cross-dressing whores in
the Meatpacking District and the homeboy queers drifting toward
the piers. Certain streets always seduced me, Eighth Avenue in the
30s and 40s for instance, one of the last safe havens for squalor
and vice in the Giuliani years, the bus station and the little hat
shops, the porn stores and the odd bodega, the sleazy dive bars
with shuttered windows where daylight was an uninvited guest;
or Avenue B with its mix of hipster joints and old-school grime,
the tapas places and the leather bars, Spanish lyrics wafting from
the tenements. I walked with the metallic tang of loneliness in the
back of my throat and the sense of seeking something, though
knowing not what it was or where to find it. I snuck glimpses into

lamplit rooms, saw men in winter sleeping over grates, all their worldly possessions contained in shopping carts. I heard laughter through open windows, saw women smoking cigarettes alone on fire escapes. Late night drew me to the sounds of jazz in those underground nests of urban cool, Small's and the Vanguard. I nursed drinks in smoky bars with the company of a pen and paper—the Cedar, the Old Town, McHale's, those dark-wood and pressed-tin interiors filled with the pleasant din of voices and the clink of ice and glass. All that love and loss on the surface of things: again and again the city broke my heart and mended it within the space of half a block, half an hour.

Once, on one of my long weekend walks, I glanced up in Chelsea and saw a strange strip of grass growing above a metal structure in the sky. Curious, I followed it—an elevated train track—several blocks north, looking for a way to scale it. I climbed a fence, scrambled up a mound of old car tires, hauled myself onto a roof, and shimmied up the iron buttresses onto the tracks. What I found floored me: a strip of meadow in the sky, a magic carpet of grasses and flowers stretching more than thirty blocks from 34th Street to Gansevoort in the Meatpacking District. Nature had reclaimed it out of view of the entire city except for a few lucky apartment dwellers adjacent to the tracks, one of whom had laid a plank from his kitchen window to allow him to step across and tend a little flower garden and a lone pine tree festooned with Christmas lights. Enchanted, I went back again and again, walking the length of the line, crawling under barricades meant to halt trespassers, always lingering in the spot where old factories and warehouses rose on either side, providing the shade for a glade of trees to flourish within site of graffitied brick. By becoming my favorite place in all of Manhattan, it showed me how badly I craved a little wildness, how starved of it I'd become. How could I sustain my attraction to the city when the thread of it I loved the most was the only place on the island abandoned by the human touch? (And not for much longer, as it turned out; in 2009 a chunk of it was tamed and turned into a scenic city park.) Ultimately, I found my instincts mirrored in a line from Thoreau: "My needle . . . always settles between west and south-southwest. The future lies that way to me, and the earth seems more unexhausted and richer on that side."

I've always liked edges, places where one thing becomes another. The railroad tracks of my youth with their remnant native grasses in the right-of-way, the New York harborside, the demarcation line of this wilderness. Transition zones, boundaries, and borderlands. I like the mixing that happens, the juxtapositions, the collisions and connections. I like the way they help me see the world from a fresh angle. Maybe it was inevitable that I'd become a searcher, a wanderer, but I think the process was helped along when my original home in the world was destroyed by the logic of industrial agriculture. On the piece of earth where I grew up not much remains to mark our efforts. A lone silo and quonset hut stand gray and forlorn against the horizon just east of the old railroad spur town of Currie, in far southwestern Minnesota. The house and garage, the grove of trees, the chicken coop, the granary, the hayloft and corncrib and farrowing barn—all are gone, burned to ash and plowed into the earth for a few more acres of tillable land, planted now with corn heavily subsidized by the government to keep it cheap and plentiful.

Sometimes I'll remember how my brother and I practiced farming in miniature from about the age of five. In a bare patch of earth next to the garage we dug little rows in the dirt, planted them with seeds snuck from the bags in the granary. Within days we had little corn plants poking toward the sunlight. For a month or so we sprayed our little field with water from the garden hose, imitating summer storms; with the toy tractors we were given at Christmas we tilled between the rows to keep the weeds at bay. I see us in our childhood naïveté, practicing for the agricultural artistry of adulthood, practicing for a future that, for reasons beyond our control, would not be ours. But before I get too carried away in sentimentality, I tell myself it was a piece of luck, our failure. It set me free to make a failure of myself in other, more interesting ways. Without that original failure perhaps I'd still be there, picking rocks in springtime, wheeling back and forth through the corn in a giant combine on a few hundred acres of someone else's rented land in autumn, playing out the final chapter in the old dream as I imagined the ways I could spend my subsidy payment. In my current line of work my government check at least feels honestly earned.

And anyway, all of that was long ago and far away. Here and now, visible twenty-five miles northwest of me over the top of the Black Range, some combination of science and art is happening, and with them something older than either. The smoke off the Diamond and Meason fires billows softly into the sky, drifting northeast. It rides the prevailing wind over the top of the Black Range, pouring into the east-side canyons and fingering toward the Sierra Cuchillos—the knife mountains—beyond the forest boundary. Come morning the air will smell sweetly of burning grass and pine duff, and my tower will cast its silhouette against a hazy salmon sunrise. No one knows exactly how the fires will roam—how many acres will burn and with what intensity; so much depends on weather—and that's a good deal of the fun. There is no formula for reclaiming fire. It may be more useful to think in terms of fire reclaiming us. Historian Stephen Pyne argues that in societies where we've locked away fire in factories and internal combustion engines, we've achieved its remystification; its uninvited appearance in nature frightens us. But if we fail to make our peace with it, fail to learn to live with it, the wildfires of the future will be like nothing we've ever seen.

THAT WILDFIRE COULD BE FUN was a fact known mainly by those who fought it. You slept outdoors. You felt a kinship with your crew members, brothers and sisters in a tough line of work. You hiked, parachuted, and rode helicopters over beautiful forests; you drove scenic roads, told dirty jokes under the stars, did your work in the wild. Though paramilitary in structure and outlook, the firefighting apparatus didn't face an enemy that screamed or bled. It wasn't quite a war; it was that slightly more benevolent thing, "the moral equivalent of war," to borrow from William James. The common endeavor protected forests and watersheds. It saved trees and homes, it guarded the public good—so the thinking went—and it was almost always a serious adrenaline rush.

But there were failures. Many became the stuff of local and even national lore, especially when drenched in death. In the northern Rockies, the big blowup of 1910: seventy-nine dead. In Montana, Mann Gulch, 1949: thirteen dead. South Canyon, Colorado, 1994: fourteen dead. These were FEAR fires, an old firefighter acronym for Fuck Everything And Run. Some who ran survived. Some did not. Suppressing fire involves a dangerous intimacy. Part of the thrill arose from the danger. That fire could be useful, even unavoidable, took far longer to appreciate.

For decades the fire of lore in this part of the world was the McKnight Fire. On the morning of June 22, 1951, a Forest Service bush plane circled the Gila on a reconnaissance flight. Despite severe drought conditions, the pilot and his spotter saw no sign of smoke, and fire lookouts reported situation normal. A couple of hours later, at 1:29 p.m., according to the *Silver City Enterprise*, "a great burst of smoke, not unlike the mushroom cloud of an atomic bomb explosion, arose over the area of the McKnight Canyon." Feeding on logging slash and thick brush, the fire burned an acre a minute for the first half hour. High winds from the southwest pushed it into standing timber. Steep canyons acted on it like a flue, and flame lengths reached 150 feet above tree line. The fire grew so fast smokejumpers were deemed useless to stop it. By daylight the next morning, 10,000 acres had gone up in smoke, and fire-line construction had barely begun.

The call went out for bodies, wherever they could be had. Men were pulled from area taverns and handed Pulaskis and shovels. Among Forest Service employees it was all hands on deck. Hopi and Navajo crews responded with numbers in the hundreds. Fort Bliss sent soldiers, and air corps students joined in from Western New Mexico University in Silver City. Volunteer ranch hands worked the fire's east side on horseback. Bulldozers blazed lines on the south and west fronts. Airplanes dropped supplies by the ton to fire camps on the Black Range crest. In the fire's second day, another 10,000 acres burned, and on the third day 10,000 more. Ash fell like snow on the town of Embree, ten miles southeast of the fire. The forest was so dry, firefighters lit back fires merely by dropping matches on the ground. One crew saw a bear running through the smoke, its fur burned off and its flesh half-charred. A reporter from the *Enterprise* captured the general mood (ellipses his): "All of New Mexico needs the one thing . . . the only thing . . . that can save the forest. It is rain. We have looked to the skies . . . now we must look to God."

By the fourth day, the fire's progress had slowed. Slop-overs spilled across lines here and there, but they were quickly attacked. Additional soldiers arrived from Fort Bliss, bringing their number to 700. Their commanders thought of it as a training mission, a chance to test them in battle. "It's the closest to combat conditions they'll find in this country," forest supervisor Ed Tucker said. Movie crews from Warner Bros. and Twentieth Century Fox arrived to shoot stock footage of smoking stumps and men smeared with soot.

One week into the fight, just as the fire appeared to be under control, the wind shifted to the north ahead of storms. Three hundred men on the south lines fled as a long wall of flame blew up at their backs. The state game department pilot, flying over the fire at the time, reported that his plane flipped upside down at the moment of the blowup, his instruments rendered useless. His spotter told a local reporter, "It seemed as though the whole front broke loose at once and swept forward. It was a horrible sight." The plume of smoke rose 18,000 feet in the air.

Soon a new fire broke out in the Gila Wilderness, a lightning strike on Little Creek. Four smokejumpers held it for a day, then high winds blew it over 5,000 acres in an afternoon. Hundreds of men left the cooling lines of the McKnight Fire, rested briefly, and joined the rush to Little Creek. The fire was eight miles from the nearest road and uphill all the way. Before it was done, it would burn 15,000 acres. It took rain and a thousand men to catch it, in that order of importance.

For the entire week the McKnight Fire burned, it shared front-page space with bulletins from the Korean War. To read those old clippings is to see just how profoundly the Cold War mentality had permeated the public attitude about fire. "Red Air Force Beaten for Fifth Time" neatly echoes the report a few days later: "Battle Against Big Forest Fire Succeeds." On July 5 a headline reported, "U.S. Battle Casualties in Korea at 78,110." A week later a similar tally appeared for the summer's big fires: "Gila Forest Loses

55,000 Acres of Timber." One reporter who obtained an aerial view of the McKnight Fire described it as a "holocaust" wreaked by "cancerous fingers of smoke." "Even at a distance," he wrote, "the thing looks menacing. It may be far-fetched, but somehow, I kept thinking of the pictures of the Bikini atom-bomb explosion and that big mushroom cloud." Talk of a "Red menace"—a popular conceit amid the ascendance of Senator Joseph McCarthy—could apply just as well to forest fires as to Asian or Soviet communists. Each threatened American prosperity. After World War II, the national forests fueled the housing boom with cheap lumber. Losing them to flames remained unacceptable. Eventually, as the Korean War wound down and the military found itself overburdened with equipment, the Forest Service would inherit its surplus aircraft: helicopters, bombers, transport and patrol planes. The tools for the one war were interchangeable with the other. In both, containment was the operative word. The enemy must not be allowed to spread.

By midcentury, a seemingly noble theory of protecting the public good had calcified into rigid dogma. It would take two more decades for the Forest Service to finally admit that fighting every fire damaged rather than protected the national forests; that instead of putting out fires, we'd merely been putting them off. Sixty years on, the McKnight Fire even begins to look like a blessing. Without it we'd have been deprived of one of the biggest aspen stands in America and much of the wildlife that love it. Its multi-level, herbaceous understory offers foraging, nesting, and denning habitat for all sorts of mammals and birds. Elk and deer browse young aspen sprouts and songbirds nest in the softwood cavities of mature trees. Though aspen tend to colonize fire-disturbed areas at high elevation—they arise in shoots from a shared root colony in what are properly called not trees but ramets—once in place they're largely burn-resistant, providing a natural barrier to fire's spread into surrounding coniferous forest. Fond of fire but resistant to fire: how's that for evolutionary genius?

ALL PAST FOLLY RECEDES, at least momentarily, beneath the roiling smokes of June. Day after day the burned acreage mounts. By the fourth of June the Meason Fire has spread across 800 acres, the Diamond Fire close to 3,000. Neither burns very hot. Each creeps through grass and duff, mostly sparing the forest overstory. Each follows the classic pattern of major Black Range fires, burning eastward up the west-facing slopes with a push from a southwest breeze.

Then, on the fifth, the wind picks up. Humidity plummets into the low teens and high single digits. By late morning both fires show increased activity, with isolated torching in the timber on their eastern fronts. The observer plane lifts off from Grant County Airport to have a look from above. The spotter recommends air support for both fires: a helicopter dropping water, a tanker dropping slurry. The fires, it is agreed, need to be slowed down.

In the early years of prescribed natural fire, the approach was to stand back and watch while a fire did its thing. It ended when it ran out of fuel or the rain put it out. If a prescribed natural fire at any point threatened lives or property, it was no longer considered a prescribed natural fire. The strategy switched to total suppression, and all necessary resources were mustered to halt it on every front. Needless to say, this either/or approach hampered the ability of crews to make a fire do what they wanted it to do. It failed to take into account that a burn might be harmless on one flank while threatening, say, private cabins on another—that a mix of approaches may be needed on any one fire.

The 1988 fires in and around Yellowstone National Park, which burned more than one million acres and cost fourteen times the park's annual operating budget to contain, were the most visible manifestation of natural burn policies that had been in place for twenty years. Though the fires may have been ecologically beneficial, even unavoidable—if not that year, some other year—the TV footage was a public-relations disaster for the Park Service. For years afterward, "prescribed natural fire" was severely curtailed in the parks and national forests, until more cogent plans could be written to prevent something on the scale of Yellowstone from happening again. The TV pictures were just too traumatic for the public to stomach, and the government land agencies had failed abysmally in educating generations of Americans raised on the propaganda of Smokey Bear.

More recently, in a program piloted on the Gila and five other national forests, fire personnel have been given greater leeway to shape burns to their liking. The Meason Fire proves a prime example of the revised approach. Prevailing wind and upslope terrain conspire to push the majority of the fire's growth to the northeast. Six miles in that direction lies sensitive trout habitat. Forest officials want to keep the Meason Fire from hitting that creek with a full head of steam, incinerating streamside vegetation and dumping ash into the water. Using the North Star Road as a fuel break, they want to check the fire on its eastern flank and force it to back slowly to the west, downslope and into the wind.

Until now, the fire has cooperated with the plan. But the sudden increase in wind speeds has flung burning embers across the road. Once they land in the grass on the far side, they start new spot fires unconnected with the main fire. These spot fires threaten to run away in precisely the manner feared. Under the old let-burn approach, fire managers would face a choice: let the fire do what it wants to do, trout stream be damned, or suppress it everywhere, halting its progress on all sides. Under the new protocol, however, they have a third option: suppress it on the east where it threatens to burn toward the trout, and let it continue on the west where no such considerations prevail. This is the option they choose.

The bureaucracy has coined a phrase for this strategy that obscures as much as it reveals. Any time a lightning fire is not immediately suppressed, it becomes, in the terms of the trade, a Wildland Fire–Use Fire Managed for Resource Benefit. In other words, where the fire is doing good, we will let it burn; where it threatens to do ill, we will check its spread.

For two days the wind threatens to whip the Meason and Diamond fires into a gallop—I measure gusts up to fifty-seven miles per hour, which means winds at twenty-five to thirty in the lowlands—and for two days the observer plane recommends fighting back with an "air show," flyboy language for helicopter bucket drops and slurry tanker runs. The forest aviation officer, who sits shotgun as the spotter on the observer plane, even goes so far as to call it "my air show," unabashed by the possessive and theatrical character of the phrase. He comes on the scene like a god from above, able to tell ground personnel what their fire looks like over thousands of acres in a matter of minutes, just by spinning a couple of donuts around it. Pretty neat trick, and useful too. For these reasons, among others, his suggestions usually hold. More often than not, his recommendation is to invite aerial company along for the ride. The bias is understandable. One would be surprised were it otherwise: imagine an air force pilot admitting army infantry were always and everywhere sufficient to the task at hand.

For all I know, every gallon of that slurry sloshed across the forest is necessary to prevent a catastrophic fire and save threatened trout. I don't get the details up here—the briefings, the maps—which is fine by me. I am both of the bureaucracy and above the bureaucracy. What I do know I piece together from static-filled radio transmissions and the occasional update given by fire personnel to the dispatcher: acreage burned, general fire behavior. Mostly, I sit and look at smoke.

So many competing interests are involved that it's always easiest to err on the side of overwhelming force and smaller, more manageable fires. There is owl and fish habitat to think about, road access to preserve, private land to protect, downwind smoke pollution to consider; there are archaeological sites to worry about, weather and bureaucratic mandates to dance with, public relations to massage. I'm glad the headaches aren't mine. Nonetheless, from my throne above the fray, I tend to think those in charge are too prone to traduce wilderness values on fires that don't threaten a thing: too quick to approve chain saw use, too quick to approve slurry drops, too quick to approve helicopter landings, complete with chain saws cutting neatly manicured landing zones—too quick at every turn to give themselves the right to do in the wilderness things the rest of us are barred from doing by law. I've seen it repeatedly over the years. The hypocrisy is glaring, though few in the firefighting establishment seem to question it. Those who fly the choppers, drop the slurry, and run the chain saws would no doubt tell you they are protecting wilderness, not violating it—that if they carve up a few pieces of the landscape with motorized tools, it's a small price to pay for protecting Gila trout, which not too long ago were an endangered species close to extinction.

If I've learned anything in my eight seasons here, it's that there are no easy answers when it comes to fire—no blanket prescriptions, no ironclad laws. Having ruthlessly suppressed every fire for seventy-five years, we created not just overgrown forests but a firefighting apparatus addicted to the big money to be made off an emergency budget line. Someday it will have to be told that the funds are not unlimited. (Do what needs to be done and send the bill later for guaranteed payment: we know how that turns out.) Just as smothering every fire the moment it's detected is no longer the answer, neither is standing back and letting every fire burn under conditions no one could construe as entirely natural. It took most of a century to create the problems we're faced with now. It will take that long or longer to burn our way out of them, and it won't always be pretty. All throughout the West fire officials, biologists, private property owners, and communities large and small are going to have to put their heads together and get creative with fire use, prescribed fire, mechanical thinning—a potpourri of approaches to the fire problem, varying from place to place and year to year as conditions dictate. Global warming won't make the task any easier, but it does make the effort necessary: we can either allow the land to burn on our terms and hope the resulting mosaic stays healthy despite rising temperatures (a big hope) or we can watch as it's reduced to blackened stumps and sterilized soil over millions of acres, the evolutionary work of millennia gone in the blink of an eye.

As much as I may quibble with specific tactics on specific fires, it's also true that no forest in America has worked harder than the Gila to bring fire back to the landscape. Statistics bear this out. In my first full season as a lookout, 260,000 acres of Forest Service lands saw fire use nationwide. Nearly three-quarters of that was on the Gila. During that summer, the Boiler Fire burned 58,000 acres beyond the northern boundary of the Aldo Leopold Wilderness, making it the largest fire-use fire ever to burn outside a wilderness area. Two years later, 290,000 acres of Forest Service lands saw fire-use fire nationwide—107,000 acres of that on the Gila. Just about any way you look at it, this forest is on the cutting edge of the effort to restore fire-adapted ecosystems. The Gila can justifiably call itself the epicenter of American wildfire.

After two days of checking the Meason Fire's northeast flank with aerial drops of water and slurry, the fire manager calls a halt. If the fire wants so badly to burn across the road and on toward South Diamond Creek, so be it. The winds have calmed somewhat, and the forecast calls for cooler, wetter weather in the days ahead. Fire honchos will sit down with maps and plot a series of what are called trigger points—plans of what to do and when to do it as the fire reaches certain predetermined landmarks. This may mean lighting a backfire to rob the main fire of continuous fuels and slow its rate of spread; it may mean employing further water drops by helicopter. It may mean calling in state Game and Fish employees to remove Gila trout for safekeeping in hatchery tanks, until the fire runs its course and the fish can be repatriated. It may even mean doing nothing at all if the fire looks good. Everything will depend on weather and fire behavior.

By the evening of June 8, the Diamond Fire covers more than 7,000 acres, the Meason Fire nearly 3,000. The mountains to my north are mantled in a haze of drift smoke stretching sixty miles. On June 9, a cold front approaches from the northeast, bringing showers and touching off lightning. Even as existing fires calm somewhat amid the higher humidity, several new ones appear, scattered across the forest and beyond it. On the morning of the tenth, John spots three fresh smokes from Cherry Mountain, two of them near Lone Mountain in the Arenas Valley. The first he calls Lonesome, the second he calls Dove, and I laugh to myself when dispatch approves his nomenclature.

In the afternoon, the fish evacuation team goes to work in

South Diamond Creek—stunning Gila trout with an electric shocker, scooping them up in nets, ferrying them by helicopter to the hatchery. How their work goes I will have to learn later from the newspapers. My boss has borrowed Mark, the relief lookout from Cherry Mountain, to spell me for a few days. Although by this point in the season I'm deeply in tune with the rhythms of solitude, the play of light on mountain landforms, the watching, the thinking, the sheer lazy lying around doing nothing, it's also true that day after day of sitting in my tower has aroused a desire to be on the ground and on the move, soaking up the particulars of all the little niches on the landscape. If I'm going to pass the torch for a few days, I may as well make an adventure of it. The Meason and Diamond fires will still be there when I return. My maps have called forth an itch it's time to scratch. The dog and I are going camping.

ALICE IS THE ONLY LIVING being I know who will take a forty-mile walk in the woods without any need for cajoling, planning, or consulting a calendar. We just go. If there's a trail on the ground she leads; if not, I make her heel while I blaze one. She carries her own food in a mini-pannier draped across her back. We find water at springs and creeks, though not all of them run in June. Only a half dozen streams can be considered perennial, and only in their headwaters, on the east side of the Black Range, and many of the springs are intermittent at best—the scarcity of water a goodly part of what makes this place so little traveled. The water I drink must first be pumped through a filter to rid it of *giardia lamblia*, a protozoan parasite found in many Western streams; it causes severe intestinal problems, or what we used to call where I grew up the green-apple quickstep.

I get a kick out of Alice on these hikes, burdened by her pack but still feeling frisky, darting off the trail to hunt small game, her attention focused on one goal, scaring up movement to then give chase. I let her lead for half an hour, let her work up a little lather. Then I remove her collar, command her to heel. With her jingling silenced and her roaming curtailed, we just might meet a bear, an elk, a deer, a bobcat, some turkeys, who knows.

As in Frisbee golf, so in hiking: the movements of my limbs help my mind move too, out of its loops and grooves and onto a plane of equipoise. I have been followed all my life, in the chaos of my thoughts, by a string of words: song lyrics, nonsense phrases, snatches of remembered conversation, their repetition a kind of manic incantation, a logorrhea in the mind, and all of them inter-mingled with sermons and soliloquies—the spontaneous talker weaving his repetitive spell. At other times, tired of the words themselves but intrigued by their internal mechanics, I find myself unconsciously counting syllables in sentences, marking each one by squeezing the toes on first my right foot, then my left, back and forth in order to discern whether the final tally is an even or an odd number. (Eighty. Even.) If I weren't a walker I suppose I'd be a television addict, a dope fiend, a social butterfly. Instead I note fluctuations on a landscape scale for hours on end each day, pow-erless to alter a thing out there in the world, utterly above it all, and when I slip into fevered solipsism I walk my way back to the place beyond words.

For an hour or two, out on the Ghost Divide, I move in thoughtlessness—allowing sights, sounds, and smells to wash over me—until I come upon a weathered human artifact. A trail sign, broken in three pieces, rests against a fallen log: "Ghost Creek 8," one chunk says, indicating eight miles between here and there. Having studied the maps, I know this is false—unless it's mile-age meant for ravens. Using the math proposed by Black Larry's

Rule No. 6, I figure it must be more like twelve miles, minimum. "Entering Black Range Primitive Area," another sign says, a relic of the time, pre-1980, when the Aldo Leopold Wilderness was only a wilderness study area and did not yet have the force of law.

Studying this second sign, I think of the men and women—many of them radical amateurs or low-paid wilderness advocates, people such as Dave Foreman and Kent Carlton and untold others whose identities remain unknown to me—who spent years of their lives inspecting the proposed boundaries of the Aldo in the 1970s, learning the lay of the land and writing letters and attending meetings and arguing with the Forest Service to include more roadless country in its final plan. Their work enlarged the wilderness by tens of thousands of acres, some of it to the east and south of where I stand. To offer but one example, the Forest Service initially argued that it didn't want the country around Apache Peak included, on the off chance it might get the itch to build a road to the lookout one day. It goes without saying that I'm glad that never happened—glad the eco-freaks, god bless their funky souls, fought back—mainly because such a road would have fragmented wildlife habitat and become yet another corridor of death and destruction through the forest, littered with empty beer cans and the roadkill carcasses of squirrels and deer. Of course I'm also glad because I doubt I could have stood to be a lookout on a mountain overrun by the motorized hordes: only three lookout peaks in the Gila still require a hike to reach them. The other seven have roads carved right to their towers.

These half-rotted trail markers attempt to impose a human scale on the land, a project that begins to appear laughably puny when geology comes into account. It may be true that I stand on the old boundary of the Black Range Primitive Area, but I also stand inside the cauldron of a great volcanic eruption. Sometime around 34 million years ago, at the beginning of the Oligocene epoch, 500 cubic miles of ash, pumice, and hot gas spewed onto the surrounding landscape, a volume roughly equal to the amount of water in Lake Ontario and Lake Erie combined. This explosion was a hundred times greater than that which gouged Oregon's Crater Lake, and more than a thousand times larger than the eruption of Mount St. Helens. The volcanic outflow—called the Kneeling Nun Tuff by geologists—was welded by tremendous heat into a layer of harder-than-nails rock that grew thinner the farther it spread from the vent where it erupted. Some of the outflow sheet traveled more than a hundred miles. Apache Peak was near the epicenter. According to Jim Swetnam, who with his wife was the lookout on Apache Peak through most of the 1980s and was there to help build the current outhouse, it took six pounds of dynamite and a lot of chiseling to carve a three-foot vault from the rock.

I try sometimes to imagine the landscape roiling, pulsing, steaming, scalding the vegetation, oozing slowly onto the surrounding plains, but the magnitude and time span of it befuddle me. Geologists say that with the magma chambers of the volcano emptied, the surface of the earth collapsed, creating an oval cauldron thirty-five miles long and fifteen miles wide. This cauldron did not stay sunken for long. Resurgent magma exerted pressure from below, uplifting chunks of Precambrian granite a billion years old, some of them overlain by sediments from a time when the area was submerged beneath a warm-water sea, as recently as 65 million years ago. Magma spurted through vents in the rock to form rhyolite domes. After cooling, contracting, and cracking, the volcanic tuff eroded here and there into pinnacles and spires. Surface water carved deep valleys and washed the sediment in broad aprons toward the Rio Grande. The result of all this vulcanism,

---

faulting, and erosion is a landscape of tremendous topographical variation: hoodoos, turrets, conical peaks, hogback ridges, steep canyons, bare cliffs. The pictorial definition of gnar-gnar.

Alice and I spend the night on the ridge, overlooking the gnarliness, and in the morning break camp and continue the walk to Ghost Creek. The trail barely shows on the ground along the last nine miles. Mostly the hike involves blind route finding, bushwhacking, retracing our steps and trying another way when the first choice fails. I lose Alice for half an hour on a massive prow of rock with couple-hundred-foot drop-offs on two sides. Once I find the path down, I shed my pack and hike back to the top to show her the way. It wasn't obvious at first, even to me, and she sometimes freezes when she gets scared. By the time we reach the creek bottom her doggy pannier is a shambles. One side of it hangs by a few measly threads. I stuff it inside my own pack and let her run free. She promptly scares up several turkeys and three cow elk.

Challenging though our journey has been, we move through the landscape in a state of awe and joy unknown to some who've come before us. A stark reminder of this fact greets us in a meadow above Ghost Creek, where fifteen graves stretch in a row beneath a tattered American flag waving pathetically from a metal pole. Nearby, tucked in an alcove of a side canyon, a plaque displays a replica Medal of Honor encased in hard plastic. These military monuments testify to the fact that control over the piece of earth where we stand was once contested in blood.

Given the fractured topography, it's little wonder the Warm Springs Apache increasingly used the area when chased by the U.S. cavalry in the late 1870s. Known by many names—Chihenne, Red Paint People, Copper Mine Apaches, Eastern Chiricahuas, Mimbres Apaches—they had hunted and camped here for hundreds

of years though they lived mostly at Ojo Caliente, in the shadow
of the San Mateo Mountains, just northeast of the Black Range.
It was an easy morning's ride from their home camp around the
warm springs to high, cool country that beckoned in the heat of
summer, beckoned for its plenitude of wild game, nuts, and ber-
ries. Given the option, they'd have preferred to stay forever on the
land surrounding the springs, with hunting privileges in the Black
Range and other nearby mountains. But the American government had other ideas.

After the end of the Civil War, settlers—many of them former Union and Confederate soldiers—streamed west in search
of land, adventure, and riches. Prospectors struck gold and silver
throughout the Gila region, ranchers undertook to meet the miners' need for meat, and the army arrived to protect the interests
of both. But the Apache had a reputation as warriors of the first
rank, and their dominion over the land around the Mogollon Plateau and the Gila River had gone largely unchallenged for centuries. They would not go quietly.

Amid the flux of new settlement, Chief Victorio and his
people lived much as they had for generations. They hunted deer,
elk, and antelope. They gathered acorns and raspberries, yucca
flowers and cactus flesh. They raided and traded, back and forth
across the Mexican border, stealing horses and cattle to swap for
guns and other provisions. In their reckoning, cattle, like deer,
were part of nature's bounty, only slower, stupider, and easier
to liberate. Horses provided both meat and locomotion. The
Apaches considered raiding an extension of hunting and gathering, an entirely natural pursuit. The ranchers they raided felt
otherwise.

As the pace of settlement increased, Victorio sensed that his
people's wide-ranging existence would be constricted. He did not

want open war with the new arrivals. He wanted to skim a bit of their bounty and he wanted the promise of a permanent reservation at Ojo Caliente. Again and again he repeated this hope. For a time his wish was granted, and an agency was established there. But government rations were insufficient and game increasingly sparse, forcing the Apaches to resume raiding surrounding ranches and villages for stock, which they then butchered or traded to Mexican settlers near the warm springs. The raiding led to pleas from the new arrivals to concentrate Victorio's band with other Apaches on a single reservation, ideally across the border in Arizona, at San Carlos. By the 1870s, whenever a horse was stolen, a cow killed, or a gun fired anywhere in southern New Mexico or northern Chihuahua, the Apaches were blamed. To this day the word *Apache* is synonymous with the practice of scalping, though it was the Mexican government that encouraged widespread use of the tactic, paying Anglo and Hispanic bounty hunters handsome rewards for Apache scalps. Cormac McCarthy's *Blood Meridian*, among the finest American novels of the past thirty years, reimagines that sordid episode with the care of a documentarian and the language of a biblical prophet; it is not for the faint of heart.

The Warm Springs Apache were rounded up and confined, at various times, to three different reservations, moving on and off them all through the 1870s. They found the San Carlos Reservation hot, dry, and dismal, with summer temperatures reaching 120 degrees; the water was bad and disease rampant. The short-lived Tularosa Reservation proved unsuitable for the kind of small-scale farming Victorio's people had practiced at Ojo Caliente; to have stayed there would have been to court starvation. The Mescalero Reservation, in south-central New Mexico, offered a more hospitable climate—the Chihenne had complained that Tularosa was too cold—but it wasn't home, and there too the government rations were insufficient to feed the population.

In 1870, the government had estimated it would cost $11,000 to buy out the mostly Mexican settlers around Ojo Caliente—none of whom had clear title to the land—and settle the Warm Springs Apache on the permanent reservation of their wish. When this idea was abandoned for good a decade later, the costs of war would run into the millions.

A series of corrupt and incompetent Indian agents inflamed distrust between Victorio's people and the government. In August 1879, Victorio abruptly left the Mescalero Reservation. He paid a brief visit to Ojo Caliente, attacked a troop of cavalry, and embarked on a raiding mission into Mexico. When he returned to Ojo Caliente, he found U.S. soldiers had sacked the camp and killed his wife along with many other women, children, and elders. This was the second of his wives to be shot while unarmed. With his future closing in upon a series of unacceptable options, each of which would force him to abandon his homeland forever, Victorio chose freedom over confinement. He and a small band of warriors stole cattle and horses, crossed and recrossed the Mexican border, and skirmished with all manner of opposition—U.S. and Mexican troops, Arizona Rangers, Texas Rangers, civilian militias—in one of the greatest guerrilla campaigns ever waged on American soil. Rival Apaches and Navajos aided the U.S. Army as scouts and lent it what little success it claimed.

Though less well known than his contemporaries Cochise and Geronimo, Victorio and his war are studied to this day by the American military. Indeed, one military historian, Kendall D. Gott, has written a monograph for the Combat Institute Studies Press at Fort Leavenworth, Kansas, in which he equates the hunt for Osama bin Laden and the wars in Iraq and Afghanistan with the battle to subdue Victorio. Published in 2004, *In Search of an Elusive Enemy: The Victorio Campaign 1879–1880* seeks to encourage our beleaguered modern fighters with tales of past glory against tough enemies:

> The Victorio Campaign bears many parallels to ongoing operations against Islamic terrorist movements. Victorio was a charismatic leader who many indeed considered a terrorist. On the other hand, his followers considered him a freedom fighter and gave him their unswerving loyalty. These warriors were fanatical in their support and willingly endured extreme hardship and depredation [*sic*] in the fight against their enemies. . . . Like today's terrorist leaders, Victorio used an international border, that between the United States and Mexico, to great effect.

On September 18, 1879, after raiding a ranch on the south end of the Black Range, Victorio and his party turned north. They picked their way through the foothills, then turned up Ghost Creek into the mountains, heading for one of Victorio's favorite camp spots. Their trail was plain—probably on purpose—and a company of Ninth Cavalry gave chase, led by Navajo scouts.

The Ninth Cavalry regiment had been mustered in 1866, in New Orleans, with most of its original members drawn from Louisiana and Kentucky. An all-black regiment, they became known to history as the Buffalo Soldiers. In the 1870s several companies of Ninth Cavalry arrived at Fort Bayard, just east of Silver City, to partake in the subjugation of the Apache. That September day in 1879, on the headwaters of Ghost Creek, marks a peculiar moment in America's westward march: black soldiers, most of them former slaves or the sons of slaves, commanded by white officers, guided by Navajo scouts, hunting down Apaches to make the region safe for Anglo and Hispanic miners and ranchers. The melting pot set to boil.

The Apache trail turned up a side canyon north of Ghost Creek. Once inside the canyon, the soldiers found themselves trapped; they'd stumbled into an ambush. Boulders and rock ledges provided cover to Apache snipers, and all day long they kept the troops pinned down. Late in the day the soldiers retreated under heavy fire, dragging away their dead and wounded. They left behind a medical wagon, fifty-three horses and mules, and most of their personal baggage. Official reports and later accounts differ on the number of casualties. Chris Adams, the Black Range District archaeologist, puts the number around five or a couple more. The row of graves is suspicious for its neatness; cavalry tended to be buried where they fell, though sometimes a later reburial consolidated scattered graves. A memorial stone counts twelve Buffalo Soldiers and three Navajo scouts among the dead. No monument commemorates the victors of the battle.

Victorio's band stayed on the move for what remained of his life. His people's rituals and ceremonies became abbreviated, perfunctory, or were abandoned altogether. They raided. They hid in the mountains. Outnumbered by as much as twenty-to-one by his pursuers, hunted by an army of men well supplied with food and ammunition, Victorio, with the help of his sister Lozen, eluded capture for more than a year even as they traveled harsh country with elders and children in tow. Few lived to tell what they said to each other at night around the fire. Victorio left behind no letters, no journals, no speeches or statements to the press. Maybe he dreamed of victory and a final, lasting peace, a return to his home at the warm springs. Maybe he suspected his resistance was a suicide mission. Maybe he spoke of an honorable death, a noble defeat. Maybe, consumed by survival in the present, he spoke not at all of the future. Maybe he knew he had no future.

In October 1880, chased from all sides by soldiers bent on his destruction, Victorio and seventy-seven of his followers were slaughtered by Mexican militias aided by Tarahumara scouts at Tres Castillos, a group of rocky, forlorn hills in northeastern Chihuahua. Some Apaches later said Victorio committed suicide rather than face the prospect of capture. The survivors were sold into slavery. The dead, including sixteen women and children, were scalped. Soldiers paraded bloody mops of hair through the streets of Mexico City to much cheering and music, then sold them to the government for a handsome reward. Victorio's alone fetched 2,000 pesos.

Alice and I linger awhile at the graves, then walk a little way up the ridge, looking for breastworks where the Apache snipers might have hidden themselves—though to my untrained eye nothing sticks out. The wind in the pines makes a noise I've heard many times in many places, but here, in this place touched by war, the gentle moan of it makes the skin on the back of my neck tingle. It is a beautiful place, but haunted by history and drenched in blood—beautiful and spooky too.

A FEW MILES UPSTREAM FROM the graves, I make camp in a meadow where two headwater streams join to form Ghost Creek. A cairn marks the spot where campers before us have warmed themselves around a fire. Their choice was sound—flat ground, ample water, plenty of firewood—so I drop my pack and lie down in the grass to rest. We have walked almost twenty miles from the lookout, and my feet are feeling it. Time to soak them for a while in the creek's cool waters.

The vegetation here shows classic ecotone mixing. The south-facing slopes reveal a desert influence, with abundant cacti and dry-land scrub. On the north slopes red-bark ponderosa far older than the Victorio war throw their shaggy shadows on the needle-cast floor. Marking the merge point, the creek meanders in a valley that now and again broadens into half-moon meadows verdant with forbs, flowers, and bunch grasses. Cliffs of pink tuff rise above the canyon. Wild turkeys roost in old-growth ponderosa, while mule deer and elk bed down on grassy benches tucked against the bluffs. Upstream, in the cool and shadowed pools of the headwaters, trout glide with a wave of their speckled tail fins.

I unzip Alice's pannier, scoop a handful of her food onto a rock at the edge of the cairn. The rat-tat-tat of kibbles on stone stirs a creature awake, a creature whose own sound is like dry corn husks shaken rhythmically: a snake. I push the stone aside with a long juniper stick. The rattler, a blacktail, coils like a lariat, its tongue testing the air. I scoop its midsection with the tip of my stick and flip the snake a few feet away, into the grass, where it quickly slithers into hiding. (No paradise is whole without the presence of the serpent.) Thankfully, Alice is off hunting; I'd meant to have a meal awaiting her return. Now the meal is scattered in the dirt around the cairn. I gather what bits of it I can, pile them on a different stone, add another half a handful.

For two days I fish upstream to a boulder-filled box of the canyon, return to nap around camp, explore the various prongs off the main stream, the hidden springs and tiny waterfalls. I fish again in the evening, downstream this time, catching foot-long trout on a bee, a nymph, a bead head woolly bugger. I pick a few wild strawberries for my morning granola, cook a late dinner under starlight. Alice joins me on my walks, but as soon as I toss a fly on the water she wanders off to do her doggy thing. She knows I'd rather she not participate, and there's plenty of adventure to be had all around—elk to roust, rabbits to chase, quail and turkey to flush. I hear her off in the distance and smile to myself as she yip-yips in pursuit of game. At night she rests beyond the glow of the campfire, ears upraised to the sounds of the forest. Her guard-dog instincts allow me a restful sleep. If anything's out there moving in the night, she'll let me know.

On the third morning I wake to haze in the canyon and a new coolness—rain on the way, real rain by the feel of it. Traveling light, I've brought a tarp instead of a tent and I don't like the thought of hunkering under it for hours on end while a storm moves overhead. I break down my camp, douse the fire, pack my food and sleeping bag, pump some water from the creek for the long hike back to the top of the Black Range. Eighteen miles to the truck. A day and a half back in town. Just long enough to sort through my mail and resupply on groceries. Then it's back to my duties as a smoke-besotted stylite.

Partway back to the crest we pass a muddy spring where a bear has wallowed not long before us. For a while thereafter the trail is wet with the bear's paw prints, the Gambel oak leaves smeared with mud. We find a tree the bear has scraped against, glossing it with a sheen of dried mud, leaving hairs stuck in the bark. I sing songs made famous by Frank Sinatra at the top of my lungs to let the bear know we're coming close behind, and though we follow in its footsteps for more than a mile we do not catch sight of it.

IT'S ALWAYS GOOD TO COME home to my cabin on the hill. Midway through my season of fire, my more permanent home in the world below begins to feel like a stopover, a destination I visit briefly to perform some minor upkeep on my shadow existence as a bill-paying, truck-driving citizen of the twenty-first century. I visit the Laundromat, send some checks to my creditors. I restock on fresh produce from my friends at the farmer's market. I exchange cash for gasoline. Then I flee once more for the hills.

Morning in Turkey Canyon is a picture of loveliness on the cusp of the solstice. Sycamores and willows shade the rock-bottom creek, and woodsmoke scents the air where families have pitched their tents in the roadside campgrounds. I drive with my window down and my arm hanging out in the coolness, the air growing cooler and cooler as we climb the winding road toward the pass. Alice stands, paces, sniffs the breeze over my shoulder, thrilled to be headed back into the wild. At Wright's Saddle I stretch my legs in preparation for the hike, gaze awhile on the dun-colored valley of the Rio Grande, beyond it the distant dome of Sierra Blanca, the southernmost peak in the United States near 12,000 feet above sea level. Then I shoulder my pack and begin the walk.

Along the way I scan the radio for news of fire. Half an inch of rain fell while I was gone, slowing the growth of the big fires, though each of them continues to smolder. The Diamond Fire has moved across 11,000 acres, the Meason close to 6,000. Typically, this time of year, managing fires that size would be a white-knuckled adventure. But the weather's been weird this season—and advantageous for fire use.

The typical weather pattern brings one-third of annual moisture in winter, followed by a dry spring that cures the grass to golden and primes the forest to burn. Scattered storms in May and June, some with dry lightning, touch off fires that in certain years burn for months, until a persistent plume of Gulf moisture smothers their spread. Old-timers place the onset of the Mexican monsoon on or around the Fourth of July, after which a couple of months of storms bring another half of the annual precipitation. The first storms of the monsoon touch off a final spasm of fires, which rarely grow large; by August, most years, the fires are finished and the creeks run high. At Gila Hot Springs, in the middle of the forest, thirteen inches of rainfall a year is the norm, on the high peaks of the Black Range more than twenty. This year a drier than normal winter prevailed, while May proved wetter than average. June has seen some mild storms with little lightning and modest rains, enough to keep the big fires from really ripping. With all due deference to a century of averages, I'm eight years into this vocation and I've yet to see two years shape up the same.

I arrive on top around noon, soaked in sweat and pleasantly exhausted, having carried with me bread, cheese, chocolate, oranges, bananas, blueberries, avocados, lettuce, onions, carrots, tomatoes, green chiles, some freshly laundered clothes, a couple of books, a couple of magazines. I gorge on lunch, replenishing the calories I've burned and then some. I nap in the tower, a deep and satisfying sleep I feel I've earned for having hiked up the peak in less than two hours. Waking midafternoon, I find a humming-bird hovering at the edge of the open tower window, shoo it away with a wave of my foot. Down the sixty-five steps, back on terra firma, I hang the sun shower from a hook on the porch and rinse the salt rimes from my skin. I eat some more. In the evening, as the sun sets in the west, I climb the tower for the reliable hour of painted pandemonium in the sky above the Mogollons. Off in the other direction, the shadow of my peak stretches in the shape of an arrowhead forty miles across the desert, the tip of it touching the Rio Grande just before sunset. In the sudden but not unpleas-ant chill of evening, I light a fire in the stone circle, stretch out next to it, watch as the nighthawks circle the meadow feeding on miller moths.

The moths represent the adult stage of the army cutworm, so named because it's been known to crawl across roads and fields in vast conglomerations. It feeds nocturnally on flower nectar in mountain meadows and hides by day in the cracks and crevices of human structures, or in among the rocks on talus slopes. It is fed upon in turn by birds, bears, and bats, which relish it for its high fat content—as much as 70 percent of the moth's body weight. Those moths that avoid their predators return to the lowlands and lay eggs in autumn. The worm emerges briefly to feed, then hibernates over the winter before feeding again in spring, in preparation for its migration to higher country.

This time of year I find the moths wedged in the window frames of my tower each morning, or in the space between my dropdown map and the ceiling. Each day I unhook the map and open the windows and watch the frenzied exodus as the moths fly away on the breeze. They invade the cabin too, emerging at dusk to beat their bodies against the windows where the moonlight filters through. When I spark the propane lamp, they attack the aluminum screen that keeps them from destroying the cloth mantle white-hot with gas. Reading after dark becomes nearly impossible, since the moths swarm both my light source and the white pages of an open book. I read a sentence or a paragraph, grapple with a moth around my eyes, read a bit more, grapple again. Beyond mid-June I tend to read by daylight and spend the evenings outdoors, around the fire and away from the silken fluttering, the moist bodies, the glowing eyes of the light-maddened moths. They like moonlight and gaslight but they show little interest in an outdoor fire.

Life quickens on the mountain in June. While it's possible, some days, to convince myself I live in a sylvan tranquility, an Eden-like innocence, all around me a predatory savagery plays out. Down at the pond one evening I see what looks like a two-headed snake, writhing upon itself like some mutant life-form. Closer inspection reveals the second head to be that of a tiger salamander. Its tail and torso already swallowed, its forelegs waving just beyond the snake's clenched jaw, it must have been caught from behind unawares. Bit by bit the snake—a Western wandering garter—walks its jaws up the length of the salamander's body, preparing to swallow it whole.

Closer to home, convergent lady beetles cluster on the trunks and lower limbs of white pines, turning them burnt-orange; on calm days they swarm the meadow to feed, slicing the air like sleet when the aphids take to flight. Short-horned lizards sit on the rocks near the outhouse, preying on passing ants. The lizards' camouflage skin, adapted to local coloring, shields them from snakes and hawks—though they will puff up like spiny blowfish or squirt blood from their eyes if threatened.

One day I kick a rotten log, testing its potential for bonfire wood. Beneath it lie dozens of tarantula hawks. I gently flip the log back into place and back away. A spooky-looking creature with an iridescent blue-black body and rust-colored wings, the tarantula hawk is a wasp that seeks out, stings, and paralyzes tarantulas to provide a host for its eggs. It drags the spider, alive but immobile, back to the spider's own burrow or some other hole. It deposits a single egg on its abdomen. When the grub emerges from the egg, it gorges on the living spider, saving the juicy organs for last, so as to prolong the life of its host and benefactor. The sting of the tarantula hawk is said to hurt more than the sting of any other insect in North America, up to three minutes of debilitating pain that feels like an electric shock to the human nervous system.

Afternoons the turkey vultures circle, indolent and bloody-headed, sniffing out the presence of death. Their arrogant flight reminds me that my time here—on this mountain, on this orb—is short. If I were to slip and fall off the lookout tower, it wouldn't be long before I passed into a new link of the food chain. A not unpleasant fate, perhaps: beats the stuffy prison of the graveyard tomb. So many dreary neighbors. So little sunlight. We're all carrion eventually, whether for birds or for worms. I'd rather my remnants soar over mountains than slither beneath sod.

But not just yet. Not this year.

On the porch, in the last of the day's light, chittering hummers visit my feeder by the dozen, greedy for my simple syrup, goosing each other with their long black bills to force a turn on one of the perches. If the goosing doesn't work, a more frontal attack, wings outspread in a posture of threat, usually suffices to free up a space. These are just the broadtails, whose jostling looks like an animated church potluck compared with the more aggressive rufous, who tend to arrive in July, buzzing ferociously and driving off all competitors before they can even reach the feeder.

Darkness ends my spectatorship of the various feeding frenzies, invites another nighttime bonfire. I gather dead pine needles and small sticks for kindling, construct a pyramid of fuel—each layer denser than the last—and light a match at its center. Soon the flames play on the tower's steel frame and throw weird shadows on the trees. Stories come to me, the voices of friends, some of them memories from nights around this very fire pit, nights of ribaldry and revelry, food and laughter, hand-rolled cigarettes, tall tales.

I think of a story Black Larry once told me—having brought to the peak with his wife Christine a feast of wine, whisky, cheese and crackers, chicken tetrazzini, chocolate cake, a pound of fresh strawberries, food we ate with joy and brio, and afterward a fire right here where I sit—how as a junior in high school he saw an installment of *Mutual of Omaha's Wild Kingdom* that changed his life. Hosted by the mustachioed wildlife lover and zoologist Marlin Perkins, the show was a television hit for nearly twenty years;

in the episode that captivated Black Larry, Perkins and his crew went to the high mountains of Chihuahua to film the world's southernmost population of grizzlies. Larry and his best friend decided that's where they wanted to go for spring break: the Sierra Madre Occidental. Larry's mother was not convinced this was a good idea. Conveniently, she played in a regular bridge game with Marlin Perkins's wife in St. Louis, Perkins at the time being the director of the St. Louis Zoo in addition to his television work. Larry's mother cagily arranged for the boys to meet Perkins and discuss their trip. They were thrilled by the chance to talk to their hero face-to-face. They went to his office one afternoon and found him seated behind a stack of maps. He showed them where his crew had flown into some remote Mexican airport, where they'd caught a bumpy overland bus, and so on. The maps were drawn on with lines and arrows and *X*s—it all seemed to Larry unfathomably exotic. Then Perkins pointed to one of the *X*s. And here, he said, here we were detained for two days by bandits.

Bandits? the boys asked.

Bandits, Perkins said. *Banditos*. They weren't *federales* but they had some pretty serious-looking guns. They took some of our film equipment. So you should prepare for similar encounters.

How do we prepare? they wanted to know.

You should probably have at least $500 cash on you, preferably in small bills, tens and twenties, Perkins said.

He went on with his account of the journey—*and here we were detained by bandits again*—until he sensed the boys had heard enough. You know, he told them, there's a place a couple hundred miles north of there in New Mexico, almost as wild, no grizzlies but plenty of black bears, elk, deer, mountain lions, bobcats . . .

The boys were sold, and in March of 1969 they began their trek up the West Fork of the Gila River. They soon realized they should have brought extra shoes for the innumerable stream crossings. That night, exhausted and soaked from the waist down, Larry cut some pine boughs to sleep on. He'd read somewhere they made a soft bed. In the morning his sleeping bag was sticky with pine pitch. They hiked that day until they saw an old ranger cabin on Blanca Creek. A wisp of smoke drifted from the chimney, beckoning them on. Inside were two Mexican cowboys who spoke not a word of English—illegals, probably, scouting for work with a ranch this side of the border. Images of *banditos* flashed in the two boys' heads. Since they could not converse, the four of them were reduced to hand gestures, a primitive sign language. One of the Mexicans showed the boys a catalog someone had left in the cabin, probably for fire-starting material. He pointed at a woman in a bra. He made a masturbatory gesture with his hand and laughed. Instead of finding this funny, the boys were terrified. Already scared of being murdered in their sleep, they now began to imagine the brutal rape they would suffer first. Each of them slept that night with a weapon secretly clutched to his breast— Larry a Bowie knife, his friend a small axe. "We were sheltered gringos from the Midwest," Black Larry admitted. "We thought if you didn't speak English you must be a criminal or a deviant." In the morning, alive but sleep-deprived, they continued their hike, over the Diablo Range in three feet of snow, nearly suffering frostbite. By the time their trip was over, Larry was hooked forever on the kicks to be had in the Big Outside. Always, though, he had to have a partner. He couldn't go alone, couldn't sleep out there by himself. Thoughts of the unknown crept up on him in the dark, kept him awake all night. With the company of one other human all that went away. Thus my friendly service to him: backcountry companion, the man whose presence in camp makes the night safe for sleep.

By midnight I've put the stories to rest once more. My mind is quiet, my fire a pulsing bed of bright red coals. The night air is too gentle for indoor sleeping. I drag my mummy bag out of the cabin, spread it on the ground, crawl inside, and commence to dream.

Even the daytime becomes a part of the dreamscape when you attain that state where you're nothing but an eyeball in tune with cloud and light, a being of pure sensation. The cumulus build, the light shifts, and in an hour—two?—you're looking at country made new. This time of year the available spectacles of lightning, hail, rain, and rainbow just continually and freakishly astound. The mountain chains stretch one in sunlight, one in shadow, one in sunlight, one in shadow all the way to the edge of the visible world. The setting sun turns the feathered drift smoke off the Diamond and Meason fires into a whorl of pink cotton candy, and despite the absence of a moon the tower casts a discernible shadow on the silhouetted trees in starlight. On stormy afternoons the lightning steps across the peaks and mesas like the quivering legs of some extraterrestrial spider. When the cumulus gather strength, I start rubbernecking, swivel-necking, watching four or more storm cells with lightning all at once. Almost every evening a rainbow appears to my east before sunset. On the twenty-fourth of June nearly an inch and a half of rain falls on the peak. All of a sudden the sky I love is gone, lost beyond the monochrome gray of a massive downpour. I'm swaddled in fog, incognito in the clouds. No point in occupying the tower; I stay in bed all day and read, rising only to make tea or heat a bowl of soup. More rains follow, a little each day, and I mark them to the hundredth of an inch in my log: .23, .04, .35, .12, .32. The days unfurl in a languor of long naps, multiple meals, a sweetly stolen indolence. After dark: moths and mist. Amid the sudden coolness my breath waves from my mouth like a frayed pennant as I sprint through the drizzle to the outhouse.

By month's end, the Meason Fire has burned just under 7,000 acres, the Diamond 17,000. Rain has slowed them both to a smolder, mostly interior smokes—downed logs, stump holes. For the first time in eight years I do not spot a single wildfire in the month of June.

4

JULY

I marvel at the calm of the Japanese haiku poets who just enjoy the passage of days and live in what they call "Do-Nothing-Huts" and are sad, then gay, then sad, then gay, like sparrows and burros and nervous American writers.

—Jack Kerouac, in a letter to John Clellon Holmes

*Visitors out of the sky * dolor & gloom, mist*
*& cigarettes * the brief but storied career of*
*Jack Kerouac, fire watcher * a call to come*
*down & a code for future lookouts * smoke*
*over Wily Canyon * high heels & firearms*
** Aldo Leopold & the fierce green fire * a*
*fawn alone in the woods * fire season lives*

ON THE SECOND OF JULY a storm comes over just past dark, lightning on all sides, some within 400 yards. Inside the cabin I cringe every few seconds as the strikes arc and strobe in the smoky-looking fog. They seem to come in a spectrum of colors, blue and pink and yellow and white, and some of them strike repeatedly in the same spot, two or three jolts to a single tree. The thunder sounds like artillery fired from the ridge tops below me. In the morning I find trees on the edge of the meadow shorn of bark in the classic corkscrew pattern, splinters the size of swords nearby on the ground—but not one inch of charred wood. Updates on the progress of the

Diamond and Meason fires have been suspended. There is no progress to report; most of the crews have been released back to station.

An oppressive melancholy sets in, gray afternoons of uselessness and curtailed views. Clouds hang in ragged clumps over the mountains, and in the mornings mist rolls out of the canyons to the east. I no longer find my sustenance in sweeping vistas but in the profusion of new color at my feet. All around a fiesta of wildflowers has burst into color: yarrow, fleabane, scarlet penstemon, skyrocket gilia, cliff primrose, Indian paintbrush, mountain wood sorrel, dozens of others. The unease, the sadness, the almost instantaneous nostalgia for events as they happen—each day on the mountain now elicits a wave of feeling centered around the knowledge that my stay here cannot last.

Most years I keep one improvement project in my hip pocket for just this moment: caulking windows, painting the tower, staining the old cabin, securing the roof against the elements. Anything to justify a few extra days or an extra tour. Eight years of such work have left the place in pretty good shape, but there's always something I can conjure. This year I've requested pickets to rebuild the fence around my propane tanks. The pickets, judged by the packers too unwieldy for the mules, require helicopter delivery, and the two choppers on the forest have been tied up for weeks on fires. But not anymore. On July 3 I get word that the helicopter is on its way. I coax the dog into the cabin to spare her the worst of the noise.

"Helicopter 305, Apache Peak."

"Apache Peak, 305, go ahead."

"Your winds are two to five, gusting to seven, out of the southwest."

"Copy that."

With a tremendous roar the whirlybird approaches the peak, circles, turns its nose into the wind. It sets down next to the tower on the landing pad, another military relic—a square sheet of Marsden matting developed for use as mobile runway matériel in World War II. Two helitack personnel duck out of the chopper and step away. When they give the all clear, the chopper lifts off for Wright's Saddle. There it will hook a net holding my pickets to a longline rope and ferry them back to the peak.

I greet my colleagues, both younger than me, one male, one female, handshakes all around. They have a bored look about them. Delivering supplies to a lookout is about the least exciting thing they've done all season—not much more than an hour's work and utterly lacking in the excitement of descending on a fire. And no hazard pay or overtime.

"Looks like it's winding down for the season," I say.

"Yeah," the guy says, "it's pretty soggy out there."

"They say we might be headed for California by the end of next week," the woman says. "If they get some fires. Things are pretty slow all over the country."

When people show up on my peak, I usually take the chance to exercise my vocal cords, revive my withered social skills, but I can't think of anything else to say, and neither can they. We fan out with rolls of bright orange tape, flagging the discreet hazards on my mountaintop for safety's sake—the clothesline, the tower's guy wires. The helitack each light a cigarette. Their posture is impeccable, their dress identical: bright yellow fireproof shirt, dark green fireproof pants. They wear sturdy utility helmets and their boots are top-notch.

Soon the helicopter is back, hovering overhead, easing the net onto the ground. One of the helitack gives the all clear over the radio, and the helicopter turns again toward Wright's Saddle, where it will touch down and remove the longline, then return to the peak to pick up the two crew members.

"Hey, man, think you can spare a smoke?" I ask the guy.

"Hell yeah," he says. "You want a few? I've got a whole pack and you're a long ways from a store."

"No," I say, "I'm pretty much quit. But every once in a while a smoke still sounds good."

He hands me two.

"Keep a spare for tomorrow or the next day."

We move the pickets onto the porch to keep them out of the rain. They roll the cargo net into a ball and cinch it tight. The helicopter returns, settles onto the pad. I tune my radio to air-to-ground.

"Helicopter 305, Apache Peak."

"Go ahead, Apache Peak."

"Thanks for the supply drop. Much appreciated."

"Any time. Just say the word."

I wave as the chopper lifts off. The pilot responds with one thumb up.

By noon both smokes are smoked, and I'm back in bed with a headache.

MORE THAN ONCE, WHEN CRAVING a cigarette here, I've thought of Jack Kerouac becoming so desperate for something to smoke that he rolled some coffee grounds early during his summer alone on Desolation Peak in 1956. I first encountered this anecdote in a book by John Suiter called *Poets on the Peaks*, a beautiful study, in prose and photographs, of the effect lookout work had on Kerouac, Gary Snyder, and their friend the poet Philip Whalen, all of whom spent at least one summer in the North Cascades in the 1950s. Having read everything they ever published about their time as lookouts—most of it pretty excellent—I was left craving more. Suiter's book clued me in to the existence of an unpublished diary Kerouac kept over his sixty-three days on Desolation. An afternoon of research revealed it was housed with Kerouac's papers in the New York Public Library's Berg Collection of English and American Literature, so one autumn I traveled east and made an appointment to see it.

On an upper floor of the library, at the end of a quiet hallway, I sat in a locked room watched over by a gimlet-eyed curator as I flipped through the Golden West notebook Kerouac had bought for nineteen cents. It was small enough to fit in a shirt pocket, and every page had been filled from top to bottom and margin to margin with Kerouac's tiny, impeccable handwriting; he'd even written on the inside of the front and back covers when all the pages were full. For a few moments I simply sat with it in front of me, marveling at its fragility. I was surprised I hadn't been ordered to handle it with gloves.

When I opened it, the first thing I saw was Kerouac's grocery list. Having made similar lists at the beginning of every fire season, I can hardly describe the thrill this gave me, the knowledge that he'd chewed Beechnut gum and eaten Hi-Ho crackers up on Desolation. He'd even used brand names! I'd expected to spend a couple of hours with the notebook, flipping through it and making some notes. Instead, over three straight days of frenzied writing spanning twenty hours—a marathon session of transcription that left me with flaring pain in my right hand—I copied the whole thing into my own notebook with a No. 2 pencil. (Pens were strictly forbidden.) Each summer now I bring it with me to the peak and read it for its talismanic power on days I feel blue.

Kerouac, at the time, was thirty-four years old. Restless and broke, he'd arrived on the mountain with two cents to his name.

Although he'd written half a dozen novels, only one—*The Town and the City*—had so far been published. *On the Road*, the book that made him famous, would not appear until the following year. Hanging out in California, becoming enamored of Buddhism, he'd heard his friends Whalen and Snyder tell colorful tales about their time as fire lookouts; Snyder recommended that Kerouac write to Washington's Mount Baker National Forest and apply for the job. Snyder's string of lookout summers had ended in 1954 when he was blacklisted by the government for his pacifist, anarchist, and pro-union sympathies.

Kerouac, torn between social and solitary impulses, had long dreamed of a mountain hermitage. In 1954, while reading the Lankavatara Sutra, a Buddhist scripture, he'd underlined the advice that "one should have his abode where one can see all things from the point of view of solitude." The frenetic cross-country trips, the hitchhiking, the freight-hopping, the drinking and drugs and parties he'd experienced in his *On the Road* years had left him hungry for peace and quiet and a place to call home. "I don't want to be a drunken hero of the generation suffering everywhere with everyone," he'd written. "I want to be a quiet saint living in a shack in solitary meditation of universal mind. . . . "

On July 5, 1956, he finally got his wish. That morning, in the Skagit River country of northwestern Washington, after a week of training with firefighters in the art of digging line and building trail, he left the Diablo guard station and floated across Ross Lake by tugboat. With the assistant district ranger and a mule packer leading the way with his supplies—including $51.13 worth of groceries he'd bought on credit—he rode the five muddy miles to Desolation in a rainstorm. On top he saw a round peak, snow in the coulees, early wildflowers; he spent his first day in classic lookout fashion, cleaning up the mess left by overwintering rats and mice.

For the first time in years he embarked on a long stretch of sobriety—completely off booze, though he did bring with him a small amount of amphetamines—and the results were not always lovely. Within the first week he would write, "Here I am on Desolation Peak not 'coming face to face with God' as I sententiously predicted, but myself, my shitty frantic screaming at bugs self—There is no God, there is no Buddha, there is nothing but just this and what name shall we give it? SHIT." All through the coming months he would alternate between states of euphoria and states of despair. "Feeling, now, happier than in years—Is it solitude or the absence of liquor?" The very next day he plunged into a funk after killing a mouse; it had raided the food basket Kerouac had rigged from the ceiling. Finding it there nibbling on green pea powder, he'd bashed it over the head with a flashlight. The murder violated his Buddhist principles and left him feeling guilty and morose.

His response to Buddhism—he was reading the Diamond Sutra almost daily—would fluctuate wildly as well. On July 12 he wrote: "I'm not going to be taken in by any ideas of transcendental compassionate communication, for it's just a nothing. . . . I'd as soon go back to Jesus & keep my mouth shut, as spin their fucking wheel (of the 'Buddhas numberless')." Three days later he wrote of meditating on a ridge, pondering "exuberant fertility and infinite potentiality" and of how he woke to "another beautiful morning above the sea of flat shining clouds." He was thrilled to see a pair of car headlights tracing a mountain road across the line in Canada. The joy, however, did not last: "Hot, miserable, locusts or plagues of insects, heat, no air, no clouds," he wrote on July 19. "For the fuck of me I'd like to get the fuck out of here—No cigarettes—NOTHING." Whatever else you may say about him, the man never left a doubt about the way he felt when he held a pencil in his hand. "I'm always deciding something & writing it down as tho it were the final word and then contradicting it later," he admitted. "I'll write it down anyway:—because it may win me."

Fourteen days into his season of solitude, he gave in to craving and radioed his boss to have a tin of tobacco delivered across the lake to the trailhead. Two days later he hiked down, spent the night on a Forest Service barge, floated around the lake while the other men fished, ate steaks with them, drank liquor with them, and, carrying a pound of tobacco and a sheaf of rolling papers under his arm, hiked back to the lookout in the morning. The company seemed to have done him good. He began writing late into the night, working on a novel about the years he'd lived in the Queens neighborhood of Ozone Park. He drank hot milk and honey in the evenings. He did push-ups and sit-ups, performed yoga head stands in the meadow. He played canasta, smoked, sang old show tunes he remembered, smoked some more, watched the northern lights over Mount Hozomeen. ("Hozomeen, Hozomeen, most mournful mountain I ever seen," he would write in *The Dharma Bums*.) He read a biography of John Barrymore. He planned and made lists, including one under the heading "List of Things for My New Life":

1. Wear soft-soled canvas shoes & slacks (chino grays) & sports shirts—& a new black leather jacket
2. Make plays for women young & old
3. Take money from homosexuals (that's a joke, son)
3A. Stand on my head
4. Drink
5. Take long walks in cities (Mexico City, Venice, Paris, London, Madrid, Dublin, Stockholm & Berlin)

6. Write the Duluoz Legend in Deliberate Prose
without stimulus, spontaneous prose with
stimulus (muscarine and benzedrine)

6A. Smoke roll-your-owns

7. Simply know that it makes no difference

7A. Write plays off tape recorder

8. Have fun

Day by day he kept a tally of the money he was making, $9.50 a shift, $350 as of August 1, money he would use for travel when his gig was up. Some days all he could do was yearn for the place he wasn't. He thought of Mexico City, where he'd stay in bed on weekend nights eating chocolate. He dreamed of Venice, good Italian food, reading outdoors along the canals. Or New York, where he'd live with his mother in the Village and work as a sportswriter. Then a storm would blow over, the looming peaks of Hozomeen would pierce the clouds, and he'd return to raving in ecstasy. He copied long transliterations of the Diamond Sutra into his journal. He sought the Void. "Wait, wait," he wrote in early August, "who wants to die in bitterness?—duck takes to water, man takes to hope . . ."

Decades before the Forest Service amended its fire policy, he perceived the futility and hubris of attacking every smoke. "As for lightning and fires," he wrote in *Desolation Angels*, "who, what American individual loses, when a forest burns, and what did nature do about it for a million years up to now?" He went the whole summer without calling in a single smoke. The one time he was the first to see a fire, his radio failed in a lightning storm. Later he ran out of batteries and had to have extras dropped by parachute. The cargo chute got hung up in a fir tree on a precipice, and Kerouac had to crawl on all fours to retrieve it. He heard a fire crew make a request for resupply—including a fifth of whisky, a case of beer, and two gallons of ice cream. The radio chatter amused him so much he copied some of it down. ("I wrote him a letter last fall and told him where I was and all"; "I'm back here in the middle of nowhere—At least I think so"; "But the Lost Creek Trail they don't believe is in existence any longer.") He found bear sign at his garbage pit, a hundred yards down the mountain—old milk cans punctured by claws. Swinging from despair to mania, he counted down the days he had left, complaining of boredom, lack of energy, lack of booze.

I wonder sometimes if, like me, he'd been given every other weekend off in town to hit the bars and mail some letters and fish the lake and check the baseball scores, or maybe, if he'd simply had an AM radio, whether he'd have thought, *Hell, I can do this—this is the sweet simple bhikkhu life*, and then come back each summer to escape the fame that threatened to swallow him after *On the Road* and maybe even live a few extra years, since, as we know, every day spent in a lookout is a day not subtracted from the sum of one's life. A period of distance from the demon lightning, a chance to clear the mind and calm the nerves. Futile speculation, I know. Presumptuous too: *If only the poor bastard had been more like me* . . . But I can't help it. His story is too sad; it begs for an ending other than the one he found, in which he drank himself to death in 1969 while living with his mother in Florida, raging in spastic bitterness.

And anyway the lookout's life was probably not for him, not for more than one summer. "I'd rather have drugs and liquor and divine visions than this empty barren fatalism on a mountaintop," he wrote toward the end of his stint. These words are especially poignant when you consider that two years earlier he'd written to Allen Ginsberg: "I have crossed the ocean of suffering and found the path at last." For Kerouac, the path of Buddhism proved too difficult, too alien to his temperament, and he eventually retreated into the mystical French Catholicism he'd known as a boy. Its fascination with the martyrdom of the Crucifixion jibed with his sense of himself as a doomed prophet destined for self-annihilation. The essential Buddhist ethic—do no violence to any living being—was a principle that tragically eluded him in his treatment of himself.

For most of his last month on Desolation he was eager to get the call that it was over. He vowed to get back to a life of adventure, complained he had nothing to write about alone on a mountain, even as he scribbled furiously. Never did man nor woman make more of sixty-three days alone on a lookout, recycling and revising the experience across multiple books and literary forms. He reworked a few observations from his journal in writing the final pages of *The Dharma Bums*, mostly passages that captured the brighter side of the experience; he wrote a calm and sentimentalized version of that summer in a reminiscence for the essay collection *Lonesome Traveler*; he went deeper and darker in the opening section of *Desolation Angels*, by far the most honest account of his lookout experience in its highs and lows, with many long passages borrowed almost verbatim from his journal; and he wrote a dozen "Desolation Blues" poems gathered posthumously in his *Book of Blues*. He also exaggerated the length of his stay to more than seventy days and fudged on the austerity of his solitude, neglecting to mention his trip down the mountain for tobacco and the resulting night of revelry on the float.

Though he was not yet known as "the avatar of the Beat Generation," he believed he would make his mountain famous, foresaw that streams of hikers would one day undertake the trek to Desolation Peak as if to a holy shrine: "Such places (where the scripture is observed), however wretched they may be, will be loved as though they were famous memorial parks and monuments to which countless pilgrims and sages will come (to Desolation Peak!) to offer homage and speeches and dedications. And over them the angels of the unborn and the angels of the dead will hover like a cloud." More than fifty years later the "little Pagoda Lookout" still hugs the top of the peak; the pilgrims come singly and in groups to see the view that Kerouac saw, sit in the shack where Kerouac sat, pace the meadow where Kerouac paced, and generally try to soak up a little of the dharma perceived to hang over the place where Kerouac tried and failed to put his demons to rest. And at least one lookout on the other end of the Rockies reads his journal sometimes late at night or on days of mist and rain, swept up in the turbulent energy of his prose, awed by the naked force of his honesty, the depth of his longing, the doomed quest of his search for lasting peace.

ON JULY 4 I SOMETIMES think I should feel a deep patriotic thrum, maybe go berserk with small incendiaries, start early on the lager; that's what my neighbors will be doing back in town, and may the fates bless them in their pursuit of happiness. But I prefer to swaddle myself in solitude and watch the fireworks forty miles away in both the east and west, blooming like tiny flowers in a sped-up time-lapse film, their elegance accentuated by the distance and the silence of their parabolic choreography. If I'm honest about it, I have to admit that my most enjoyable national holidays have occurred in the company of friends and loved ones: the summer Martha and I recorded twenty minutes of ourselves on a cassette tape, *ooh*ing and *aah*ing in the tower at the fireworks shows in Elephant Butte and Silver City, twenty straight minutes of only the two words *ooh* and *aah*, varying their length and tone to impart gradations of meaning, breaking down from time to time in wheezy belly laughter; or the year Mandijane and Sebastian showed up with wine and steaks and some kind of hybrid of bottle rocket and Roman candle that we aimed at the tower from a hundred feet away, contra Forest Service rules and regulations. Frivolity and nonsense ought to be a part of anyone's pursuit of happiness, and they've certainly made for my most memorable Fourths of July.

Lucky for me the holiday falls roughly midway in my annual twenty-week hiatus from rather too obsessively following the folly and farce of what passes for our politics. By a quirk of schedule—the shape of fire season in the American Southwest—I feel, by the first week of July, almost nothing but love for a country that would produce even one human being with the idea to preserve a forested commons from the onrush of our most destructive tools. Most of our best ideas have enlarged our definition of small-*d* democracy: one human, one vote; public schools for all our children (desegrated, of course); the creation of what the great Bob Marshall, founder of the Wilderness Society, called "the people's forests." Yes. To America, then—to our forests, our Founding Fathers, our Bettering Mothers, our magnificent flora and fauna diverse and healthy in its native element, of thee I proudly sing. Alone, this year. And thank goodness. I do not sing well. A fine reason to keep the song short.

JOHN ON CHERRY MOUNTAIN, HEDGE on Monument Mountain, and Sara on Snow Peak each spot new smokes on July 6—tiny fires, little snags and logs on the ground, all of them smaller than a tenth of an acre and none of them a threat to get big. Lightning hits all around me, but no smokes show in my corner of the forest. Hail bounces in the grass like bingo balls. Rain collects in pools amid the rocks in the meadow. Alice can't sit still. She paces the cabin, a pensive look in her eyes. She curls up on the floor at my feet for a while, rises and returns to her bed, then comes to my side again when lightning stabs a couple hundred yards away, the thunder like some celestial gong crashed above our heads. I roll a new sheet of paper in the typewriter and stare at it for several minutes, my attention repeatedly disrupted by the flash, the crack, the boom of the rapid-fire strikes. All of a sudden the hair on my arms stands up, a flickering penumbra of yellow-white light surrounds us, and I feel a percussive blast almost before I hear the sound: the cabin's been struck by lightning. Thank goodness the thing is grounded. For several minutes there's a weird smell in the air, like an overheated radiator, and my heart jiggles in my chest like a fox in a burlap sack. Lightning continues to pound all around, and I count off the distance of the strikes from the peak—five seconds between flash and thunder equals one mile, and I rarely get to five. Alice, pressed against my leg for some small sense of comfort, shudders every time the thunder sounds.

Late in the afternoon the clouds break. I climb the tower for a gander at the country. The sky is immense and cleansed by rain, the earth below it a palette of muted blues and greens and browns. With the dust washed from the air, the vistas boggle the mind, my horizon stretching as much as 200 miles away. In all my seasons I've never seen the view so clear, so I open my notebook and begin to name and count the visible mountain ranges—the Wahoos, the Datils, the Cuchillos, the San Mateos, the Magdalenas, the Fra Cristobal Range, the Oscuras, the Caballos, the San Andres Mountains, the Sacramentos, the Organs, the Franklins, the Doña Anas and the Rough and Ready Hills, the Sierra de las Uvas, the Good Sight Mountains, the Potrillos, the Mimbres Mountains, the Cookes Range, the Floridas and the Tres Hermanas, the Cedars, the Big and Little Hatchets, the Animas Mountains, the Pyramids, the Peloncillos and Chiricahuas and Big Burros, the Pinaleños, the Silver City Range, the Pinos Altos Range, the Diablos, the Jerkies, the Mogollons—more than thirty in all and me in the middle of them, goggle-eyed and rapturous, alone in my aerie in the vastness. Caged by glass but caressed by sky, I come as near as I'm able to a perception of the numinous. The writer Richard Manning has argued that "the most destructive force in the American West is its commanding views, because they foster the illusion that we command." I can't say I've ever felt that way here. If anything, the views on offer command me: *sit and be silent.*

My moment of enchantment is broken by a burst from the radio, my boss calling.

"Apache Peak, Division 62."

"Division 62, Apache Peak."

"Hey, Bubba, how's it goin' up there?"

"Not bad, chief. Just sitting here counting mountain ranges."

"Copy that. Good day for it, I'm guessing. I've got some news for you if you're ready to copy."

"Let 'er rip, Skip."

The news is unwelcome. Dennis doesn't have much use for me anymore, not with the rains coming almost daily. Unless the weather pattern changes, I'll be granted four more days on my mountain and that will be that for the season. We lookouts are getting the hook.

As consolation, he offers an extra week of work in the office, helping to write an updated lookout manual for future rookies. The prospect does not entice me. I can't imagine writing a document bland enough to earn a stamp of approval from a U.S. government agency. Inspired by Kerouac's list making and Black Larry's Rules for Black Range Travel, I've been working all summer on a code for lookoutry, my best advice for all who come after, though I know it's not what Dennis has in mind for his manual:

1. Do not miss a chance to nap.
2. Leave the place better than you found it.
3. Never piss into the wind.
4. Go buck naked in the tower now and then for kicks.
5. Learn what it means to ride the lightning.
6. Cut a good supply of wood for the start of next year.
7. Feed the hummingbirds.
8. Have a hobby: reading, knitting, playing the ukelele. Something.
9. Sleep outside when the weather permits.
10. Love your neighbor as yourself. (Lacking human neighbors, love the bobcats and the turkeys, the chipmunks and the tassel-eared squirrels.)

If a week in the office turns out to be mandatory, I'll suck it up and do my time; I was once conscripted for worse. During one of my first summers on lookout, an old grazing-allotment fence was torched in a fire on the north end of the district, and after I was pulled from the tower I was ordered to join the crew rebuilding it. Thus was I thrust into company after a season of solitude: a gang of three cowboys who packed supplies for the fence and meals for the crew, and the crew itself, an ever-shifting group of tough young firefighters living in tents both acrid from camp smoke and richly rotten from dirty work socks. The cowboys were a colorful bunch. Like *vaqueros* of old, they drifted from job to job, working on a series of ranches, rounding up renegade cattle in the wilderness, packing with mules for forest work crews, guiding hunting parties in the fall. They spent their downtime back at the Pine Knot Bar, otherwise known as their "headquarters," and around the campfire their stories concerned variations on only two subjects: whisky and pussy. The liveliest of their tales arose from the confluence of the two, as if they were rivers on whose banks all good cowboys waited as life's treasure flowed past for the taking. After I shared with them some of my own whisky supply, they decided my company was tolerable, though I couldn't say the same for the work, which I hope and pray I'm never forced to do again.

The job was not only the worst kind of grunt work, it was counterproductive: cutting down old barbed wire, rolling it up, pounding new fence posts, stringing and stretching new wire—all in rocky, steep, undulating mountain terrain—so a rancher could resume running cattle on the public domain at below-market lease rates, his efforts subsidized by my mindless labor. It's one thing to rail against government farm subsidies; it's quite another to suffer the humiliation of *being* the subsidy. Cattle grazing and fire suppression have been the banes of this country. Why the Forest Service would correct one component of a warped cycle, by letting a fire burn where it wanted to, while perpetuating another, by refitting the land for grazing, is a contradiction that bedevils public-lands management in this part of the world. The argument in favor of grazing boils down to sustaining a 120-year tradition. Whether such a tradition is worth sustaining despite the mounting costs—fiscal and ecological—is a whole separate question the agency prefers to duck.

I'm not unsympathetic to the notion of preserving rural traditions. I grew up in the bosom of traditions that somehow felt eternal, on a farm where I became intimate with domesticated animals from the moment I could walk; I knew how to castrate a pig before I knew how to read. We lived a mile down the road from the homestead my great-great-grandfather staked out in 1887, in the Des Moines River country of southern Minnesota. My great-grandfather lived there all his life. He saw both the first motorcar and the last train to come barreling over the hill out of the east. My grandmother and her siblings retrieved corncobs from the pigpens to burn in the stove and warm the house during the Depression. My father and his cousins held late-summer hog roasts and potluck socials. We were regulars at 4-H club meetings. Church attendance was nonnegotiable, as was respect for all manner of Jell-O salad cuisine.

In truth, that way of life, which felt timeless to my child's mind, was barely more than a century old, and its establishment required just as much butchery as the subjugation of the Apache, except in our part of the world it was the Sioux who had to be done away with. For years my father's cousins struggled to drain and farm a swampy spot on the prairie called Slaughter Slough, the name of which gives a flavor of how the so-called Sioux Uprising played out. The farmers who, unlike my father, survived the lean years of the 1980s have been propped up by government subsidies for their commodity crops, sometimes to the tune of millions of dollars. In the rural county where I grew up, the federal government paid a quarter of a billion dollars in farm subsidies from 1995 to 2009. The cheap corn grown there is shipped west and south to the Great Plains, or to Texas or California, where it's used to fatten hogs and cattle on industrial feedlots whose tremendous odor and death-camp architecture testify to the ecological depravity of the entire enterprise. All the corn syrup guzzled down the gullets of America's overweight children, all the ethanol being distilled in heartland refineries, all of it underwritten by as wasteful a government spending program as now exists this side of the defense industry: these things do little to reassure me that precious rural folkways are being preserved in the place I once called home. An analogous situation holds true for public-lands ranching, an anachronism sustained by government predator-control programs, giveaway rates on public lease allotments, stock ponds gouged by bulldozers to catch rainwater, barbed wire strung through the mountains on the taxpayers' dime—all of it calculated to favor one domesticated animal we can eat over the diverse array of life we do not.

WITH MY DEPARTURE IMMINENT, I draw up measurements for my new picket fence. The project serves not only to keep me here a few extra days, it distracts me from feeling sorry for myself. I'm not ready to leave, but then I never am when the word comes up it's time to go. I always want one more fire, one more week, one more tour, one more month. Each year I've settled instead for the prospect of one more season, which forces me to improvise a winter scramble after a paycheck at a job I can abandon without remorse come spring.

I try not to think ahead just yet. Morning dawns clear and warm, a few puffy clouds in the sky, one of those Black Range summer days of unsurpassed splendor—dappling shadow, lazy light, bees buzzing drunk with nectar. Good weather for the work that awaits me. I balance the pickets on the sawhorse and brush them with a weatherproof sealant. I cut some lengths of two-by-six with a handsaw for the perpendicular braces, stain them too. This takes most of a day. Every hour or so I break for a glass of water in the shade of a white pine tree. Around four o'clock I look up from my work and glance off toward the Pinos Altos Range in the west. There, on a west-facing slope twenty miles away, a brand-new smoke slithers into the air, a single snag by the looks of it, visible for who knows how long. A sleeper smoke from the previous day's storms, nothing to be worried about given how wet the forest is—but I'd hate to be scooped by another lookout.

I drop my paintbrush and hustle toward the foot of the tower, radio in hand. I churn up the steps, pop through the trap door. I spin the sight on the firefinder, nail down an azimuth—265 degrees, fifteen minutes—and make the call to Mark, the relief lookout on Cherry Mountain.

"Cherry Mountain, Apache Peak."

"Apache Peak, Cherry Mountain."

"Hey Mark, I'm picking up a new fire a few miles due east of you. Could be hidden by a ridge from your view, but you may see a bit of drift smoke if the light hits it right."

"Copy that. Let me have a look."

Three minutes later he calls back.

"Phil, I can't seem to see anything out there. Can you tell me if it's in a canyon or on a ridge top, maybe what aspect it's on?"

I lift my binoculars for a magnified look. The radio squawks, Mark again, and this time his voice is shot through with a childlike glee: "Phil, I got it now!"

"Copy that. Let me know your azimuth when you have it. Mine's two-sixty-five and fifteen."

Sara on Snow Peak, having overheard our conversation, chimes in with her azimuth too, so we're able to pinpoint it to the quarter section, high on a ridge above Wily Canyon, three and a half miles east of Cherry Mountain. I encourage Mark to call it in. Generally we lookouts ascribe to the rule of finders keepers—you see it first, you get to call dispatch—but this time I make an exception, since the fire's so close to Mark and he knows the country there better than I do. Plus it's on the Silver City District, not the Black Range District, meaning Mark will be the one to name it no matter who calls dispatch. (A quirk of protocol on the Gila: if a lookout from one district spots a smoke on another district, the thrill of naming is denied the spotter.) He calls it the Wily Fire, and during his rundown he graciously tells the dispatcher I saw it first. A little later the dispatcher calls back to say the fire won't be fought. For the time being it will remain in monitor status, and Mark and I are to watch the smoke and report any major changes.

The smoke sputters for two days, two days of warmth and sun and not a drop of rain on the forest. In the mornings and again in the late afternoons I give dispatch a smoke-volume update as the fire punks around in the soggy duff, doing its best to stay alive. Its continuing survival is my last hope for reprieve. In the meantime I finish work on my fence, take down the clothesline, sweep and mop the cabin. I cut more wood and stash a supply for next year in the tack shed. I roll up my maps and slip them into their protective tube. I zip my Olivetti Lettera 26 into its carrying case, the sleek machine a gift from a friend who won it in a raffle in Tuscaloosa, Alabama, and told me the rumor was it had once been owned by the writer Barry Hannah. I figure I can hump out all my belongings in two trips down the mountain with a full pack each time: I'll hike out for days off this weekend and return next week for whatever's left.

Just as I'm locking the cabin to leave for the hike to the pass, I hear Dennis calling on the radio.

"Hey, Phil, am I right today's your hike-out day?"

"Yep, I'm just about to head down."

"How's that Wily Fire looking?"

"It's hangin' in there, hoss. Smoke volume's pretty decent in the late afternoons. If I had to guess I'd say it's closing in on an acre."

"Copy that, bubba. I've been looking at the forecast and it looks pretty dry for the next few days at least. Monsoon's not setting up like we'd hoped. I'm thinking we ought to keep the towers staffed but we don't have relief for your days off. How soon can you be back?"

"Sooner'n you can say Sourdough Mountain."

AFTER THAT FIRST CALL FROM Dennis, telling me to prepare myself to leave, I'd been searching for consolations to the season ending so abruptly. Among them were visions of a romantic night out for me and Martha, to celebrate our wedding anniversary—a reunion by candlelight at Diane's restaurant in downtown Silver City, a bottle of good wine, and later a dance or two in the space between the pool tables at the Buffalo Bar while the jukebox played our favorites. We've not seen each other in more than a month; she extended her stay in Massachusetts because her father's cancer had returned, and despite some early treatments the prognosis isn't rosy. Back in town for a quick resupply, I tell her I'm sorry—but duty calls, and the season can't last forever. She shrugs. "What can you do? It's the U.S. Forest Circus. So we'll spend the day together on the peak." She's been down this road before, and compared with her father's condition it's a very minor inconvenience.

William Kittredge, one of the writers I admire most, once wrote: "Back in my more scattered days there was a time when I decided the solution to all life's miseries would begin with marrying a nurse. Cool hands and commiseration. She would be a second-generation Swedish girl who left the family farm in North Dakota to live a new life in Denver, her hair would be long and silvery blonde, and she would smile every time she saw me and always be after me to get out of the house and go have a glass of beer with my buckaroo cronies." Martha and I like to laugh about this, in part because of its self-mocking wit but also because she's tolerant of far greater eccentricities than her husband's need for an occasional beer with his buckaroo cronies.

She begins work at the local hospital this month, her first job in nursing after three years of school. She's understandably nervous. Three nights a week, from 7:00 p.m. to 7:00 a.m., she'll care for a half dozen sick and largely helpless people, many of them in pain, some of them on their lonely way to death. Her clinical rotations have already revealed that a goodly portion of her work will involve an intimate acquaintance with the products of the human bowel. If I were the man who'd just crapped in his robe at four in the morning, confused, embarrassed, no longer in control of my bodily functions—and aren't we all that man in the end?—I'd want her as my nurse. She's empathetic and efficient and not at all squeamish, the ideal blend of traits in her chosen profession.

On the morning of our anniversary I cook her breakfast— bacon and eggs, a rare treat on the mountain, packed in this time by Martha. While she takes the dog for a long hike, I roam the meadow picking wildflowers, which I arrange in a glass of water on the table in the cabin. In the early afternoon we play a few games of cribbage in the tower. She scratches my back. I rub lotion on her feet. For a while she reads a magazine while I glass the country for smokes. Then she drops through the hatch, but not before issuing a stern warning: "Stay in the tower until I tell you to come down." She's cooking dinner and wants it to remain a surprise.

I wonder sometimes if my loner tendencies make her life more difficult than it needs to be. We married because our minds are enriched and our senses sharpened in each other's presence, and because our interests dovetail nicely—but my attraction to solitude precludes time together and doesn't really dovetail with anything but its own perpetuation. For most of the year we like to read in bed next to each other, cook together, share music we've discovered, drink wine and dream up the ways our story might turn out. I lead her on hikes to backcountry hot springs she'd never have thought to visit on her own. She introduces me to movies I missed in a farm boy's youth devoid of access to decent film. I continue to argue for my summers here on the basis of necessity—for my mental health and for my creative life. The days and weeks alone help me focus my all-too-often scattered thoughts, and I like to think I leave each summer a calmer, happier person. She likes to visit the mountain, likes the time we spend together here, but there's not enough of it, not with her responsibilities back in town, and I know she'd prefer it if I gave up the job. A third of the year is a long time to be away.

The sun lowers in the sky, smoke drifts lazily over the mountains—the Meason and Diamond fires have both come gently back to life amid a run of dry weather, burning once more in the grass and duff, a few dozen acres a day—and I feel a tremendous peace come over me. My dog naps in the meadow. My wife cooks dinner in the cabin. I've been on the clock all day, a professional watcher of mountains. These are the moments I hope to hold close when I'm old and palsied, ransacking my memories for succor and solace.

When I hear her call, I descend the tower to find Martha in a floral-printed dress and high heels, pulling a pan of lasagna and a loaf of garlic bread from the oven. There's a fresh salad in a bowl on the table, an open bottle of wine, and two fancy wineglasses she carried up the hill for the moment. I smile and lean near to kiss her.

"I don't know how I got so lucky."

"I don't either," she says. "Now let's eat."

. . .

AFTER DINNER, IT'S MY TURN to entertain.

Back in springtime I lugged my friend Jaxon's 12-gauge shot-gun up the hill. One night in town I'd casually mentioned to him my interest in hunting turkey, and he'd offered me the use of his Remington 11-87. He assured me he didn't use it much anymore. If I bagged a bird, he wanted one breast saved for him. But the howling winds of late April and early May—spring turkey season in New Mexico—made most mornings unsuitable for calling in birds, and my enthusiasm for killing the creatures with whom I share the mountain turned out to be less avid than I'd imagined. For most of the summer the gun has rested in the corner of the cabin, unloaded and untouched.

Recently, while boxing up trash left by past lookouts to send out on mules someday, I came upon the black-and-white TV once watched here by a lookout for whom the birds and the clouds weren't enough. During my first season it remained operational, wired to a primitive antenna and capable of drawing a signal for the local public station and a couple of the networks. I carried it out of the cabin one afternoon and smashed the screen with the axe end of a Pulaski, the classic double-bitted firefighting tool, axe on one side, mattock on the other. Now it's time to finish the job.

"I think you might want to change into something more com-fortable for this," I tell Martha. "Especially the shoes."

"Good," she says. "I'm beginning to feel ridiculous."

She steps into a pair of Carhartt work pants, laces up her hik-ing boots. I load the gun. We call Alice into the cabin to keep her out of our way. Together we traipse through the greening meadow, down to the corral, where the TV sits with some other junk on the seat of an outhouse abandoned half a century ago because the vault was full. The wooden structure leans to one side, battered for decades by the elements. I set the TV on a tree stump and pace off fifty steps. Time for target practice.

Guns were not a part of Martha's New England upbringing; until this day she's never fired one. I can see from a slight posture of fear that she's still not sure she wants to. Nonetheless I go over its mechanics, how you load it, how you fire it, where you find the safety switch, and how you toggle it one way or the other. I demonstrate how you hold the weapon cradled above the armpit, suggest she cover her ears lest the sound of it startle her. I point the barrel, ease the safety switch off. I fire a shot that grazes the TV's yellow plastic frame.

"I don't know about this," she says.

I flip the safety on, hand her the gun.

"Just hold it for a while," I say. "Get a feel for it."

"I didn't think it would be this heavy."

For her first time shooting I'd prefer to have something small and sleek, maybe a .22 pistol like the one Mandijane carried when she worked here, a gun that nestled neatly in the palm of the hand. But for years we've been too poor for me to justify spending money on guns—not that I was all that keen on them anyway, not after my younger brother, alone in his Albuquerque apartment on a Sunday night in June 1996, put an SKS semiautomatic rifle to his temple and fired two rounds through his brain. He was twenty-two years old. His married girlfriend had broken up with him the day before. His blood-alcohol content was slightly more than the legal limit for drivers in the state of New Mexico. These facts hinted at an explanation for his death but they did no more than hint. In beginning my search for answers, I read the police and autopsy reports and requested the crime-scene photos taken by the cops when they entered his apartment. The only thing that made any sense was the gun. Some people use a version of it to hunt deer, but I'd never known my brother to hunt deer. The SKS is a military assault rifle of Soviet provenance used by the armies of certain eastern bloc countries. It was meant for killing people and it had done its job.

More than anything it was the context of my brother's gun that spooked me, its presence in the police photos, cradled in his cold and mottled arms, his head exploded from the force of the bullets. For a long time afterward a morbid voice inside my skull thought that to have a gun around was to court the possibility of killing myself, if only to know for an instant what he knew in his final moments, when he pressed the barrel against his temple, tempted by that one small squeeze of the trigger that offered eternal escape. For me to spend a whole summer in the presence of a gun has been a kind of test. Long periods, days or even weeks, have passed without my thinking about it, which I take as a hopeful sign—that I can live with it nearby, much less handle it without a fear that I'll instinctively turn it on myself. To share it now with Martha, to introduce her to the pleasures of shooting target under circumstances so incongruous—blasting away at a television set to enliven the celebration of our anniversary—seems to me, in my residual Catholic-boy fondness for symbolic gestures, the final step in putting my revulsion of guns forever to rest. I know Martha senses this without my saying it. I poured all my feelings about my brother's death into those letters we shared my first full summer here, when I wrote to her at length about his sad and lonely end, how a poisonous guilt gnawed at me for years, how I clawed my way out of it by sleuthing through his past for the sources of his sadness.

She aims, slides the safety switch off, squeezes the trigger. The gun bucks in her arms; the empty shell twirls into the air. Her shot misses high, peppering a shrub behind the TV.

"Holy cats," she says. "My ears are ringing. And I think I bruised my shoulder."

"Let me see."

She opens the neck of her shirt. The print of the gun butt is visible against the flesh below her collarbone. I trace the raised and reddened outline of it with my finger.

"That looks painful."

"I'll be all right. I just want another shot at that TV."

BY THE THIRD WEEK OF JULY, with the days calm and sunny, the Wily Fire burns along at a modest pace, now 35 acres. I give smoke updates to dispatch twice a day, and the other lookouts do the same for various fires around the forest: the Scott Fire (three acres), the Turkey Fire (220 acres), the Moore Fire (1,700 acres). Every morning I discover a new wildflower in bloom, and beneath the densest forest, where the earth remains moist, mushrooms have begun to sprout.

Martha spends three days on the mountain, three days of delicious domesticity. We bake peanut butter cookies and banana bread. We take warm baths in the tin tub, scrub each other's backs with soapy water. We sit on the porch watching the riot of hummingbirds swarming the feeder, thrilled to discover among them a single calliope—the smallest breeding bird in North America. Although the rufous hummers have arrived to stake their claim, they've yet to intimidate their rivals, the broadtails, into moving on: some days three dozen birds buzz the porch at once. In the tower one afternoon we see a storm hammer the wilderness to the northwest, and after it clears a new smoke appears. I call John at Cherry Mountain for a cross. We triangulate our azimuths and place the fire above Salt Creek. The Salt Fire: a single snag burning feebly on a south-facing slope. It's so far away—twenty miles—and so small it's barely visible to the naked eye; it takes me a minute to explain to Martha where it is. She's seen her share of fires on her visits here and is unimpressed by this one. By the next morning, when a helicopter goes looking for it, it will be dead out.

After dark Martha and I sit around the bonfire in council. Our lives are about to change, now that she'll be a nurse and not a student. For starters, we'll get our dental work done on this side of the Mexican border, and with any luck I won't have to return to the trenches of the wine and spirits industry, hustling for tips and swapping quips with the tipsy. The size of her paycheck will determine my options. No matter what, we decide, we're retiring the Shaggin' Wagon, the maroon Dodge Caravan that no longer runs in reverse. Mere mention of it triggers a story.

"Remember that time I came to pick you up at the pass and you weren't there?" she asks. I do. This was in the days when Martha used to shuttle me to the trailhead and then return to pick me up for my days off. We had our rendezvous carefully planned, but once when the relief lookout hiked in a few hours early I decided to hit the trail and spare Martha some of the drive. At the pass I hitched a ride sixteen miles out of the mountains to the Hot Springs turnoff. There I would wait until I saw her coming, flag her down as she passed. After about forty-five minutes I saw the van coming down the hill into the valley. I stood on the shoulder next to my pack. As she drew near I waved a couple of times, thinking she'd slow as she saw me. But she didn't slow, and my waves grew frantic, and then she was upon me and I saw, as she passed, that she was fiddling with something on the seat beside her and wasn't looking my way at all. Then I was staring at the back of the van as it motored through the valley and up into the foothills, aware I'd made a huge mistake.

Three hours, a long walk, and a hitched ride later, I was back at the pass, back where I'd started. I walked up the short dirt road to the supply shack at the trailhead. Martha was nowhere to be found. I could see her tire tracks where she'd parked and waited and finally turned around and left, and there was a note stuck to the door of the shack saying she'd been there at four o'clock. I was glad she hadn't started hiking up the trail in search of me. She must have gone down the other side of the mountains into Gaylord to use the phone.

I sat at the pass and waited. Another half an hour passed. By now it was almost seven o'clock, and I was worried she was rounding up a manhunt to come in search of me. I heard a car coming so I jumped back into the road. Tourists, it turned out, a couple from Vermont. They said they'd give me a lift down to Gaylord, so I could look for Martha there. We got half a mile down the road when we saw the van coming toward us. They flashed their lights, pulled over. Martha took one look at me getting out of the backseat and shook her head. Thinking I would save her an hour of driving, I'd cost her three hours of searching. She kept shaking her head the whole time I explained what happened.

"I thought maybe you broke your ankle hiking out," she said.

"Nope, I'm just a dumb shit," I said.

"Can't argue with you there."

Now, thankfully, we can laugh about it, and our laughter leads to still other stories of the past that distract us for a while from our future uncertainties, but eventually our conversation turns to the subject of her father and his health. His final days may be upon us; any moment now Martha could get the call from her brothers telling her it's time to come home. She will want to be there if worse comes to worst, doing what she can to ease his passage. That will mean asking for indefinite leave

or giving up her job entirely, potentially kicking the breadwin-
ning back to me. I'll go with her to Massachusetts and find
a job or help her family around the house; I'll stay in New
Mexico and tend the hearth with the dog until she returns.
Whatever she needs.

I sit behind her as she sits Indian style, my arms wrapped around
her, both of us silent, eyes drawn to the mesmerizing flames of the
bonfire, the night around us cool and calm, the dog curled up in the
meadow nearby, the stars gone crazy in a moonless sky, our future
unknown but our present no way other than we'd wish it.

ALONE AGAIN, I WALK the next evening with the dog down the
southwest slope of the peak, hoping to forestall the deep blue funk
that always comes upon me in the hours after the departure of my
closest friend in the world. Watching Alice hunt as she wanders to
and fro off the trail lifts me out of my interior dramas, places me
back in the sensual world of squirrel tail and bird wing, color and
movement, shadow and light.

A mile down the trail she stops, ears upraised, tail suddenly
still. I crouch behind her for a view from her level, but before I'm
able to see what's got her attention she's off at a dead sprint, bark-
ing and yipping. Out ahead of her I hear the low moan of a cow,
and another, and another.

I hustle after but I can't keep up. I hear her barks echoing
down the mountain as she chases the cattle into the upper reaches
of Apache Creek. Soon I come upon evidence that several cattle
have been munching on the grass, their piles of dung still wet.
They've strayed several miles out of their owner's allotment on
the forest to the west, probably through a broken fence. These
four-legged locusts, with their shit-smeared rumps and moon-eyed
stares, flies orbiting each of their orifices, have done more than

anything else to inflict widespread damage on the public lands of the American West. Yet given the power and persistence of the cattlemen's lobby, they continue to graze on the public domain, trampling riparian areas, hastening erosion, pulverizing wildlife habitat, disturbing the fire regime, and generally wreaking havoc on the land wherever they roam.

The faithful insist this sort of public-lands ranching is "a way of life," but for certain species in these parts it's more accurate to call it a way of death. For thousands of years the Gila country was home to grizzly bears, but no more. Hide hunters, cattlemen, and their mercenary hired guns took care of that. Once, too, the Mexican gray wolf thrived here, but ranchers in partnership with government exterminators wiped them out north of the Mexican border. There was a time when such developments were trumpeted under the banner of human progress, but Aldo Leopold helped us see them for the tragedy they were. "A thing is right when it tends to preserve the integrity, stability, and beauty of the biotic community," he wrote, elucidating his land ethic. "It is wrong when it tends otherwise." By this measure, public-lands ranching in the West has been, for the most part, deeply wrong.

Eighty years ago the decline in wolves led directly to the biggest permanent wound ever inflicted on the Gila Wilderness: the construction of the North Star Road. Many justifications were made for this breach of Leopold's roadless reserve: improved communications between remote ranger stations, quicker response time by firefighters, easier access for ranchers. Some also argued the scenery wasn't that great anyway—who cared about dry and dusty piñon-juniper steppe? But the major clamor for it came from deer hunters too tender to stray from their automobiles in the pursuit of big game. They and the Forest Service had their eyes on an alarming explosion in the deer population along Black Canyon Creek and the East Fork of the Gila River. Deer had multiplied beyond the carrying capacity of the land because large predators—principally the Mexican gray wolf—had been poisoned, trapped, and shot to near extinction, in part to make the country cozier for cattle. In his early days as a forest ranger in Arizona, Leopold himself had gunned down a wolf, an incident about which he writes with rueful poignancy in *A Sand County Almanac:*

> We were eating lunch on a high rimrock, at the foot of which a turbulent river elbowed its way. We saw what we thought was a doe fording the torrent, her breast awash in white water. When she climbed the bank toward us and shook out her tail, we realized our error: it was a wolf. A half dozen others, evidently grown pups, sprang from the willows and all joined in a welcoming melee of wagging tails and playful maulings. What was literally a pile of wolves writhed and tumbled in the center of an open flat at the foot of our rimrock.
>
> In those days we had never heard of passing up a chance to kill a wolf. In a second we were pumping lead into the pack, but with more excitement than accuracy: how to aim a steep downhill shot is always confusing. When our rifles were empty, the old wolf was down, and a pup was dragging a leg into impassable slide-rocks.
>
> We reached the old wolf in time to watch a fierce green fire dying in her eyes. I realized then, and have known ever since, that there was something new to me in those eyes—something known only to her and to the mountain. I was young then, and full of trigger itch; I thought that because fewer wolves meant more deer, that no wolves would mean hunters' paradise. But after seeing the green fire die, I sensed that neither the wolf nor the mountain agreed with such a view.

By 1919, ten years after the fateful encounter, only a dozen wolves remained in New Mexico, and in Arizona they were gone altogether, with terrible consequences for the health of the land. Without predators to cull the sick and the weak, herd size increased and deer starved from lack of forage. Late in life, Leopold was haunted by his early advocacy of killing predators to protect cattle and expand game herds. About that road which sheared off one-third of his original roadless preserve—severing the Black Range from the rest of the wilderness to the west—Leopold would eventually concede, "I was hoist of my own petard."

Leopold didn't live to see the Mexican gray wolf listed under the Endangered Species Act, in 1976. Soon thereafter, U.S. and Mexican officials began a cooperative program to breed the wolves in captivity, having captured five survivors deep in the Sierra Madre. In 1998 eleven wolves were released into the Blue Range Recovery Area, just across the state line from the Gila. More wolves were released in subsequent years. The U.S. Fish and Wildlife Service had hoped to see a population of a hundred healthy wolves in the Gila and Apache-Sitgreaves national forests by the end of 2008. During a 2009 survey their numbers stood at fewer than half that. By all accounts the recovery program has been an abysmal failure. Many ranchers think the subspecies should have been allowed to die out or remain in zoos; conservationists had hoped for a bigger, healthier population of wild wolves.

The wolves' tenuous toehold has been a symbolic blow to a ranching culture long buttressed by a sense of entitlement and a righteous nostalgia for a time when the national forests were carved into a series of private ranching fiefdoms and anything that stood in the way could be crushed by political power or simply shot on sight. To the eyes of an outsider it verges on the tragic to see members of an iconic American livelihood reduced to curdled bitterness and benighted propaganda in their public statements, even if the iconography never did square with reality. When you drive around the Gila you're bound to encounter homemade placards warning you that wolves are on the prowl nearby, hungry for your pets and even your children, never mind that wild wolf attacks on humans are practically nonexistent in the historical record. (By contrast, between 2006 and 2008, eighty-eight people died from dog attacks in the United States, and more than a thousand Americans visit the emergency room each day for dog bites.) The most vociferous voices in the ranching community claim the wolf-recovery program is a government conspiracy to run them out of business, even though more than 95 percent of their livestock losses result from other reasons—lightning strikes, falls off cliffs, mountain lion predations, disease—and Defenders of Wildlife runs a compensation trust that pays ranchers full market value for a confirmed wolf kill of a cow. Some modern cattlemen proudly proclaim their descent from the pioneers who settled the region in the face of renegade Apaches and cold-blooded criminals, while at the same time they complain to the local newspapers about the psychological trauma and sleepless nights they and their children suffer after hearing a wolf howl. They're either the toughest people on earth or the most timorous, depending on the proximity of *Canis lupus baileyi*.

The wolves have not been helped by their designation as an "experimental, non-essential" species, which denies them full protection under the Endangered Species Act. They've tended to stray outside the man-made political boundary meant to encompass their approved habitat, and when this happened they were either rounded up and rereleased inside the recovery area or trapped and locked away in captivity, depending on how often they'd strayed. If a wolf killed three cows in a calendar year, it was in turn hunted and shot by government marksmen for doing a thing that made sense—eating the slowest, dumbest, meatiest thing around, an interloper poorly adapted for survival in a wild landscape. The wolves, too, may prove susceptible to the rigors of the wild, their genetics concentrated among so few founders. And despite the prospect of a year in jail and a $50,000 fine for anyone found guilty of killing an endangered species, more than two dozen wolves have been poached by persons unknown.

Leopold's story of killing a wolf, a deep parable about good intentions and unforeseen consequences, has yet to be absorbed by those who still view predators as vermin rather than as creatures of power and majesty—an inextricable presence in a healthy biotic community. Remove them and the weave of life in a place begins to fray: we know this now, in part because Leopold learned it the hard way.

He had his final revelation on the subject of Southwestern land health during two trips to Mexico in 1936 and 1937. By then he'd long been settled into a professorship of wildlife management at the University of Wisconsin. Along the Rio Gavilan in the Sierra Madre Occidental, while hunting deer with a bow and arrow, he wandered through grass-rich hills of live oak and wide-open stands of ponderosa pine, marveling at the beauty and integrity of the landscape. Here was a place untouched by cattle, where a healthy population of deer coexisted with their natural predators—wolves and mountain lions—and fires burned freely over the mountains with "no ill effects." He saw no evidence of overbrowsing. Erosion too was nonexistent. The contrast with the mountains of Arizona and New Mexico came as a shock: Leopold realized that in all his years as a forester he'd never seen truly healthy land. "Our own Southwest was pretty badly misused before the idea of conservation was born," he wrote in his essay

"Conservationist in Mexico." "As a result, our own conservation program for the region has been in a sense a post-mortem cure."

The postmortem cures continue to this day. Gila trout, once on the brink of extinction, are thriving in the headwater creeks of the Gila River. Cattle grazing has been reduced in riparian areas, bringing increased stability to the watershed. Wolves again howl on this side of the border, and the free-burning wildfire Leopold once disparaged as "Piute forestry" has been allowed to roam on its native terrain. With every passing decade the Gila continues a process of rewilding and renewal—tentatively, uncertainly. Yet the paradox inherent in the concept of "managed wilderness" is as stark as ever. Which fires are good and which are bad? Can wild wolves and domesticated cattle coexist on public lands? Does poisoning streams of non-native fish to restore native species have consequences for amphibian and invertebrate life that we can't yet comprehend? These questions remain unresolved in our time, the politics surrounding them as prickly and polarizing as ever.

As for parables of good intentions and unforeseen consequences, they are not the province solely of scholars and the famous dead. One late-July evening, Alice and I hit the trail for an evening hike. We are a couple hundred yards below the tower when we come upon a tiny fawn alone by the side of the trail, in a little clearing. It doesn't move—its legs are tucked beneath its quivering belly—but it appears alert, and frightened by the dog's curiosity; it makes a little mewling sound like a newborn kitten. I keep Alice at bay while I gingerly inspect the fawn for wounds. I find none. For reasons more instinctual than intellectual, I assume it's been abandoned—alone and helpless in an exposed place, right next to the trail—so I cradle it in my arms to carry it back to the peak, thinking I might find a way to save it. It kicks its hind legs and bolts off. It runs perhaps thirty feet before its legs give out and it collapses, whining and crying. Alice gives chase and has to be restrained again. This time the fawn does not resist when I pick it up, and we march up the hill, the dog circling us and jumping until I order her ahead of us on the trail. Shyly she obeys, and we march like some strange, tragicomic parade through the woods to the meadow on top. There must be something about the fawn's coat that doesn't jibe with my sinuses; by the time I reach the cabin I'm sneezing violently, on the verge of an asthma attack, and frightening the poor creature even more.

I set it up in a bed of sweaters and jackets in the tin bathtub. I heat some soy milk in a pan on the stove; I manage, by squeezing its jaw and tilting back its neck, to force it to swallow a little milk spooned onto its tongue. I'm scared I might make it choke, but with a little practice I become reasonably adept at making it take the milk.

I radio the Embree work station and ask the guy there to call Gila Wildlife Rescue, to see if I can get some guidance on how to feed and handle the fawn. I know the guy who runs GWR. In fact I'd just been at his house a couple of weekends earlier, where he showed me an owl, two kit foxes, and a fawn he was nursing back to health. Unfortunately the listing for him in the phone book is out of date, so the guy in Embree calls and leaves a message with the state Game and Fish Department. After a couple of hours they call back to say they'll send an agent out of Las Cruces to pick up the fawn. I'm surprised they don't have someone closer but relieved at least that help is on the way.

"Do they know it's a five-mile walk to get here?" I ask.

"I'm not sure," comes the reply. "I'll check."

Ten minutes later he radios back. "Game and Fish says to return the fawn to the place where you found it, leave it there, and let nature take its course." What I am really being told, gingerly but unmistakably, is that it's my fault for interfering in the first place. I've disrupted the natural order of things. I should undo my error, return it to the place of its discovery, forget about it.

I begin to entertain a host of troubling questions: Have I doomed it merely by touching it? Will the scent of me, the scent of a human, ensure its rejection by its mother no matter what? Will its immune system have any defense for the microbes I carry? Is my own stupid sentimentality the real cause of its doom? Have I—and the dog—frightened off its mother, a doe who would've returned when we passed? All of a sudden I realize I don't know the first thing about the nursing habits of newborn mule deer. What had looked so jarring to me—a little fawn alone in the woods, defenseless against all who would harm it—may have been an utterly natural occurrence. I feel the warm flush of shame at the depth of my ignorance.

At dawn the next day the fawn is no longer in the old tin tub. I find it looking lifeless, curled under one of the bunks in the corner of the cabin, an unmoving little coil. I feel its side: warm to the touch. I heat more soy milk and forced some down its gullet with a spoon.

In the afternoon I rouse it and carry it outside, place it on the ground. It wobbles and sways but remains upright. I walk a little ways off. The fawn begins to follow. Every moment afoot its legs appear stronger, more sure of each step, until I jog a few paces and it breaks into an awkward lope, hind legs splayed to maintain a precarious balance. I lead it around the meadow for a hundred yards, the fawn following like the obedient pet I do not want it to become.

When we stop, the fawn moves between my legs, its muzzle in the air, searching for a nipple. What, I think, do I have that resembles a nipple? A saline spray bottle, a nasal moistener. I empty the bottle, rinse it, and realize I've run out of soy milk. I remember some powdered milk, years old, stowed in the back of the pantry. Insufficient, no doubt, but it's all I've got. I heat water and add powder and suction the mixture into the spray bottle. I hold the bottle between my legs like a nipple. To the fawn it does not feel like a nipple, of course, but after several false starts and milk dripping down its chin it finally understands the mechanics and suckles for a few seconds on the nozzle, and I deliver perhaps half an ounce to its shriveled stomach.

Since it seems strong, I decide now is the time to return it to the place I found it, if return it I must. When we reach the scene of our first meeting I sit on a log and wait to see if it will recognize the place. It comes toward me, jaws in search of a nipple once more, its cries louder, and all of a sudden I begin to weep. The fawn wanders unsteadily along the slope below the trail, seems to sniff at the earth in recognition, in a way I've never seen it sniff. It walks a few feet and curls next to a log, and there I leave it. It does not stir as I walk away up the hill.

In the night I come as close as I've ever come to prayer. More hope than prayer, in the end, but fervent in the way we think of the desperate and prayerful. I wish for the fawn to be wild, to run in high mountain meadows under moonlight, to feel the cold splash of crossing a creek in autumn. To know desire, pleasure, pain. To at least be given a chance. A life. Not merely birth and death.

Even more I wish I'd never seen it.

The next morning I amble into the meadow to piss and I hear its faint cries. By walking in ever larger concentric circles I discover it lying next to a rock near a salt lick visited by its own kind, mature mule deer, whose droppings litter the ground there. It had walked up hill god knows when. I did not hear it in the night.

When it sees me—or maybe when it hears me, for its eyesight appears poor—it comes to me with its muzzle more insistent than ever for a nipple, its teeth gripping the inseam of my pants and gnawing, sucking. Again I fill the saline spray bottle with milk and get a little down its throat.

I call the dispatcher on the radio, in one last effort to enlist help. I tell him to try anything to reach my friend at Gila Wildlife Rescue, to let him know I can either meet him at the pass or drive to town with the fawn. I will find a way to carry it the five miles down. Then the district office comes over the radio and says: Game and Fish instructed that you take it away from the tower, do not touch it, and let nature take its course.

"Roger that," I say, helpless and sick with remorse.

That afternoon the fawn curls up beneath the sawhorse, ten feet from the cabin door, the place it seems to feel offers it a modicum of safety as its strength wanes.

Alice is lovely with it, resisting her urge to sniff at it, paw at it, make it a plaything. The fawn even comes to Alice and crouches beneath her in search of a teat, and Alice simply stands with her legs splayed and looks at me, uncomprehending and perhaps a little scared. Much of the day she barks into the trees at the edge of the meadow, barks seemingly at nothing, at phantoms. As if to say: Do not come near the fawn. Do not prey on the defenseless.

My radio remains quiet. I sit on the porch and watch the fawn's torso rise and fall with each breath. In its last moments of strength the fawn lies on its side and gallops in place, all four legs churning like pistons, like a dream of running. Its cries grow louder, higher pitched, more insistent. Then all of a sudden it is still.

To bury it or leave it as carrion would seem a desecration—either one the act of a conscienceless murderer—so I build a pyre. I've been saving dry wood for weeks under a tarp for late-season bonfires. I burn one batch of wood down to a flaming bed of red-hot coals and then I place the fawn on the bed and cover it with yet more wood. I stack it as high as the stone circle will allow. The flames leap and the wood crackles. Before the fire burns completely down a deluge pours from the sky and reduces it to a core of molten coals and a drift of blue-gray smoke. It rains a quarter inch in half an hour. The meadow is perfumed with the smoke, and for days afterward the memory of that smell will haunt my every waking hour.

July ends with midnight lightning over the Black Range, anvil-headed cumulus glowing in the moonlight. Two days pass before the sleeper smokes show; in the span of an afternoon I call in three new fires. John at Cherry Mountain spots two more. The observer plane manages to find a fire deep in a canyon ten miles south of me, where smoke would have to rise 500 feet for me to see it—as it eventually does, three hours later. On the last day of July ten active fires are visible from my vantage: the Diamond, the Circle Seven, the Powderhorn, the Outlaw, the White, the Trigger, the Wily, the Hightower, the Rainy, the Thompson. The forecast calls for a spell of hot, dry weather ahead—high pressure building over Arizona. Fire season lives.

5

AUGUST

I follow the scent of falling rain
And head for the place where it is darkest
I follow the lightning
And draw near to the place where it strikes

—Navajo chant

Fire myths ancient & modern ∗ a memory of smoke in lower Manhattan ∗ last fires & lazy days of rain ∗ waking above the clouds ∗ elk in the meadow ∗ a hidden cache of curiosities ∗ the consolation of words & the escape from words ∗ questions for lookouts past

IN THE ORAL TRADITIONS of Native Americans, fire often comes to us as a gift from the nonhuman world. The Nez Perce tell of the time a beaver hid below a riverbank near a stand of pines. Alone among all living beings in having access to the power of fire, the pines warmed themselves around it as they met in council. A single ember from their bonfire rolled down the bank and was captured by the beaver, who held it to his breast and ran. For a hundred miles he was chased by the trees but he couldn't be caught, and along his journey he gave fire to the willows and birches and many other trees. Ever after, any human in need of fire obtained it by rubbing together the wood of the trees where the fire was stored.

In the stories of the Cherokee, fire came to earth when lightning struck a sycamore tree on an island. All the animals could see the smoke across the water; they met to plan a way to get the fire off the island. Raven volunteered to fly across and bring the fire back; when he landed on the tree his wings were scorched, and ever after he was black. The screech owl tried next, but a blast of heat and smoke nearly burned out her eyes when she looked down into the tree, and ever after her eyes were red. Many others tried unsuccessfully, until the water spider came up with a plan: she would skitter over to the island, spin her thread into a bowl, place an ember in the bowl on her back, and recross to the mainland. When she returned, she shared fire with all the animals and humans.

In similar stories from other cultures, teamwork is often required among the animals. They take fire from its original owner, a species of tree or animal or even a band of humans that previously held a monopoly on its use. An arduous journey is usually required, as are cunning and stealth. Sometimes the animals pass fire one to the other as if in a relay race; they conspire to spread fire far and wide among their fellows, littering the land with it as they fly and run. Often the ultimate beneficiary is humankind, for whom the acquisition of fire represents a great leap forward. In many myths, humans only become fully human once they possess fire.

In modern times, our culture's most powerful fire myth is printed in bold letters beneath the picture of a friendly-looking bear, and the motifs are neatly inverted. Instead of fire coming to humans as a gift from the plant and animal worlds, fire comes to plants and animals as a scourge, and only the intervention of humans can prevent catastrophe.

Smokey Bear first appeared in the 1940s, a creation of the Wartime Advertising Council. After a Japanese submarine shelled an oil field on the California coast in 1942, not far from the Los Padres National Forest, the government feared that incendiary weapons could touch off fires in the woods along the Pacific. With the number of wildland firefighters depleted by the war effort, the Forest Service sought to involve the civilian public in a campaign to prevent wildfires. The Ad Council coined several slogans— "Forest Fires Aid the Enemy"; "Our Carelessness, Their Secret Weapon"—and early posters in the campaign showed an image of Bambi. But Walt Disney authorized use of the fawn for only one year, and when the license was up, the Ad Council and the Forest Service invented Smokey Bear.

In 1950, the agency found a real-life mascot. While fighting the Capitan Gap Fire in New Mexico's Lincoln National Forest, a group of firefighters retreated from a blowup and took shelter on a rock slide, covering their faces with wet handkerchiefs to avoid being burned. When the smoke cleared, the only living thing in sight aside from themselves was a bear cub clinging to a charred tree, its paws and hind legs singed by flames. The firefighters rescued the bear, its burns were treated, and afterward it was sent to live out its life in the National Zoo in Washington, D.C. The little cub was an instant media sensation, a living emblem of the pernicious evil of wildfire and the benevolent hand of man. Smokey's admonition—"Only YOU Can Prevent Forest Fires"—would become one of the most recognizable slogans in the history of advertising, his friendly mug inseparable from the idea that fire had no place in a healthy forest. Schoolchildren were urged to write letters to Smokey after hearing his message in speeches and videos, and they responded in such numbers that Smokey was assigned his own zip code. His presence nowadays is more muted, his message updated with modern lingo: "Get Your Smokey On." Were he given the space needed to articulate it, his more honest assessment might read something like: "Remember—Only YOU Can Prevent Your Cigarette or Campfire from Starting a Wildfire We Are Forced By Long-standing Protocol to Suppress with Every Available Resource so as Not to Encourage Promiscuous Pyromaniacs; On the Other Hand Some Fires Started by Lightning Ought to Be Allowed to Run Their Course, for Reasons of Forest Health and Ecological Renewal—Fires We Call Wildland Fire-Use Fires Managed for Resource Benefit . . . "

The area encompassing the McKnight Fire, for instance, considered at the time a catastrophe beyond all others in the annals of Southwestern wildfire, now represents some of the finest bear habitat in the Black Range. I see evidence of their presence all over—tracks in the pond muck, scat on the ground, overturned rocks where they've scraped for worms and grubs, trees they've rubbed against leaving telltale hairs behind, some of them brown, some of them red, some of them almost blonde. Acorns and raspberries—their abundance a result of fire opening the conifer canopy and exposing the earth to sunlight—entice them in the late summer.

One evening I return from a hike with Alice and sit with her in the meadow, scratching her belly and whispering nonsense in her ear. On the edge of my vision I see something small and black and fuzzy moving. I heed my instinct to hustle Alice into the cabin. I step back outside and tiptoe toward the bonfire ring, moving in the direction I saw the bear—a cub, it turns out, maybe sixty pounds, about the size of Alice. It ambles along the tree line on the peak's east slope, nosing along, briefly unaware that someone's watching. I pause and remember mama must be near. The cub's furry little head appears around the edge of a rock outcropping; it lifts its snout in my direction. Maybe it smelled me. Maybe it saw the blur of my movement. Alerted to potential danger, it turns and trundles downhill, *Ursus americanus*, king of these woods and cohabitant with fire.

WE TOO WERE ONCE COHABITANTS with open flame. We let it loose on the land to reshape vegetation to our liking, preparing the land for agriculture. We used it to make our food more palatable. It lit our camps at night and held the mysterious dark at bay. Now, of course, it is unwelcome in the places where most of us live. An urban people, we are ever more removed from its workings. We do the vast majority of our burning now in secret. The flame has been hidden in internal combustion. We burn fossilized vegetation that did not burn however many million years ago; we suck it from the seabeds and drill it from the plains and mountains, we refine it, we burn it, and mostly we don't even see the smoke. The by-products of this burning linger in our atmosphere and seem likely— by warming the average temperature of the surface of the earth— to exacerbate the drying of the planet's flora, increasing the occurrence of catastrophic wildfires. When fire makes an appearance in our cities nowadays, the word we often use for it is *terror*.

Ironically, after all my years as a lookout, the one big fire I've seen up close—so close I inhaled its visible particulates—was a fire of the kind that exploded our invisible burning into a horrific tableau. One bright morning in September 2001 I received a call from a friend in New York who knew I lived without a television. She told me, in a voice wracked with panic, that the World Trade Center towers had been hit by airplanes. I put on my suit jacket, left my apartment, and ran to take a subway to my job at the *Wall Street Journal*. I was on journalistic autopilot: the biggest story in the world was happening right across the street from my workplace, and therefore I had a professional obligation to get there, even if

I usually copyedited pieces about theater and books. I strongly suspected my superiors would find a better use for me that day.

Partway to the office, my train stalled and didn't move for an hour and a half. Since we were stuck underground, we had no way of knowing the severity of the situation downtown, and when at last we were discharged from the train at Union Square, I continued the journey to the office on foot. In Chinatown, the police had cordoned off the streets. No one was allowed any farther. The towers, in the distance, were swathed in a cloud of black smoke; still stuck in a news vacuum, my mind couldn't comprehend that they were no longer standing. I did know that if I was intent on getting to the office, I had but one choice. I would have to reenter the subway system and walk through the tunnel.

The entrance to the Franklin Street station was blocked with yellow police tape. I looked at the campaign posters for the mayoral primary—due to take place that day—taped to the railing above the stairs and thought that if I crossed the police line there would be no one to rely on thereafter but myself. What else was I going to do? Go back to my apartment and listen to the radio? Sit in a bar and watch TV? I lifted the tape, descended the stairs, and, in a last gesture toward civilized norms, swiped my MetroCard instead of jumping the turnstile.

No trains were running. No clerk was in the token booth. I waited a few moments to see if a train or an MTA worker would appear, but there was only an otherworldly quiet. With no one around to stop me, I lowered myself onto the tracks and began walking through the tunnel, creeping through the dark, careful to avoid touching the third rail. Not even the squeak of a rat marred the silence. It would be the only time I ever heard nothing in New York.

Ten blocks later, when I emerged into the light of the Chambers Street station, the platform was coated in dust, and ahead in the tunnel I heard water rushing with a sound like a waterfall. A couple of cops were in the station, hanging around the token booth, their radios occasionally squawking. I waited until they wandered off and then I climbed the stairs to the street.

I emerged a couple of blocks north of the towers, or at least where the towers had been. The streets were covered in ash and office paper. A cop stood alone in the middle of the street, watching a burning building, which I later learned was 7 World Trade Center. I walked over and stood next to her, both of us mesmerized. After a couple of minutes she looked at me. "That building's probably going to go," she said. "You might want to get out of here." She didn't order me to leave. She seemed to assume I wouldn't. She merely offered it as a suggestion, one among a series of options available to me, take it or leave it.

I picked up a discarded dust mask, put it on my face, and began to make my way around the smoking rubble, through streets flooded with greenish-yellow water, or ankle-deep in fine gray powder. After crossing the West Side Highway, I entered the World Financial Center complex. The Winter Garden's glass roof was shattered in places, and the palm trees in the courtyard were pallid with ash. All the shops were empty. I climbed the emergency fire stairs in 1 World Financial Center. I saw no one. The office had long been evacuated and was now, at least on the floor where I worked, coated in a thin gritty film blown in through shattered windows, though the computers still ran on the power of a backup generator. It was one of the most unnerving moments of my life, standing in that empty newsroom, wondering where everyone was, hoping none of my colleagues had been hurt or killed, all those computers humming with no one in front of them.

I went to my cubicle, blew the ash off my keyboard, set a newspaper over the dust on my chair, and logged on to my computer. I sent an e-mail message to the group of colleagues on my wing of the paper, asking if anyone needed anything, since I'd made it to the office. Those equipped with laptops immediately wrote back and told me I was crazy, that I ought to get the hell out as soon as possible, there was nothing I could do for them there, a gas line might explode, the building might collapse. I logged off and walked around the office, inspecting the damage, hoping I might see another editor, but I couldn't find a soul. I circled back to my desk. The telephone rang. It sounded a little forlorn, even spooky, amid the unusual silence of the newsroom. I picked it up. It was my mother calling from Texas, where she was on vacation with my father, watching TV with her in-laws. I could tell from her voice that she was frightened witless. I said I was fine, we were just now evacuating the building, all was well, I would call her later in the afternoon. I hung up and checked my voice mail. There were eight frantic messages from friends wondering if I was okay. I got up and went to the men's room. I felt strangely reverent as I stood before the urinal, aware I'd be the last man to piss there that day, that week, perhaps even that month or longer. (Almost a year, as it turned out.) The irony, when I thought about it later, was vertiginous: I had less devotion to the idea of the paper than anyone else I knew there, yet I'd risked my safety to get to the office—and for nothing. I was useless. Little did I know that if I'd wanted to be of help, I should have hopped a ferry to New Jersey, where a small group of editors was putting together a paper that would win a Pulitzer Prize for spot news coverage. The *Wall Street Journal* of September 12, 2001, carried a banner headline in letters nearly as big as the masthead: "TERRORISTS DESTROY WORLD TRADE CENTER, HIT PENTAGON IN RAID WITH HIJACKED JETS."

I suppose I could tell you how the smoke smelled when I went back outside, like every kind of noxious chemical burning you've ever known mixed into a cloud so thick you could almost chew it. I suppose I could tell you how, if you looked up at the bright blue sky a certain way, you could see waves of tiny glass crystals floating and sparkling like iridescent sea anemones. I could describe the firefighters standing around in the smoke and dust, holding their heads in their hands, some of them openly weeping, aware that hundreds of their colleagues were dead. But many people have written about what they saw that day, and I have nothing new to add. I was just one of the couple dozen spectators at the edge of the rubble, vainly hoping for a call to join a rescue operation, snapping pictures with a digital camera I'd snatched from the office, as if to preserve, in some form outside of myself, the ghastly images searing themselves on my brain—images eight seasons of wildfire have yet to put to rest.

This, I'm afraid, is another version of the fires that will plague us in the future.

IN THE FIRST WEEK OF AUGUST I wake each morning to a world bathed in smoke, the contours of the land softened by the haze. More than half a dozen fires still burn to my north and west, all of them exhibiting only mild activity. Overnight the drift settles in the low places as a layer of cool air flows downslope from the mountains, drawing the smoke down with it. Daytime heating produces gentle breezes and thermal currents that allow the smoke to lift once more. The Meason Fire holds steady at 7,055 acres, burning only in the interior. The Diamond Fire continues at a modest pace, edging beyond 20,000 acres—a crew of seven monitors its growth, mapping the one small length of active perimeter and collecting field data. The Wily moseys along, burning a few dozen acres a day, now 270 acres in all. The Turkey (430 acres), Trigger (390 acres), Cougar (150 acres), and Hightower (140 acres) each do their thing, growing slightly every day, firefighters camped nearby to monitor fire behavior. Other, smaller burns find wetter conditions or sparser fuels and black-line themselves at a few dozen acres, no one having viewed them up close, the grueling hikes required to reach them judged not worth the bother. (The smokejumpers are long gone, shipped off to California and the northern Rockies as fire season has marched inexorably north.) A crew rides horseback into the Rainy Fire, initially thought to be a promising candidate for fire use. The crew is so deep in a canyon they have no radio contact with anyone but me, so I relay their supply orders and fire updates to the dispatcher. After four days of scouting the country, assessing fuels and working trails as potential outer boundaries, the crew gives up when the fire fizzles at the foot of some bluffs, having burned less than two acres. They ride nine miles back to their truck at the trailhead and immediately get a call to help on a new smoke that threatens some radar towers over near the Arizona line. The last big suppression fire of the season on the Gila, the Radar Fire will burn 367 acres before firefighters corral it.

On August 6 wicked storms light up the sky all over the forest, and when the rains clear, I call in what will turn out to be *my* last smoke of the season—a little puffer near the crest of the Black Range on the upper end of Lost Canyon. Surrounded by aspen, the Lost Fire burns a single snag down to its roots overnight and is never seen again.

With the fire danger diminished, I get a full four days off for one of the few times all summer, and when I return on August 12, I'm told this will be my last ten-day hitch.

The season has come full circle. I revert to my April routine, climbing the tower a few times a day for a quick look around. The

234

rest of the time I nap, bake cookies, take long walks with the dog. We visit the pond, which is filling up again after going dry in June. We look in on the hidden sheepherder's shack where a frying pan still hangs on a nail inside the door, though no one's lived here in sixty years and the roof has caved in. We seek out wild raspberry patches where the last of the year's fruit is turning ripe; I pluck a handful to accompany my evening treat of chocolate, leaving the rest to the bears. On days of heavy rain, hail drumming on the metal roof, I cloister myself in the cabin, drink hot tea, read in my sleeping bag with a fire going in the woodstove. Tattered flags of fog drift past the mountain when the rain breaks. Sometimes I pause in my reading, copy a line in the commonplace book I keep:

Multitude, solitude: identical terms, and interchangeable by the active and fertile poet.

—Baudelaire

The idea of the contented hermit who lives close to nature, cultivates his garden and his bees, is trusted by animals and loves all of creation, is some kind of archetype. We think we could be like that ourselves if somehow things were different.

—Isabel Colgate

A man is rich in proportion to the number of things which he can afford to let alone.

—Henry David Thoreau

Walking is the great adventure, the first meditation, a practice of heartiness and soul primary to humankind. Walking is the exact balance of spirit and humility. Out walking, one notices where there is food.

—Gary Snyder

I wake one morning to find myself alone above the clouds—pure blue skies overhead but below me what looks like a vast ice sheet stretching in all directions, the whole world white and sparkling in the sun, blindingly radiant, the peak rising up like an island in a glacier. Then the storms come once more, the fog moves in, and in the evening the lights of all the distant towns are lost to me. Cut off from all evidence of human settlement, alone in the starless dark, I light the propane lamp and sit down to write last letters to friends. I tell of the armored stink beetles taking shelter from the rains in the outhouse, and of how the mule deer now sport healthy russet coats, much different from their ragged, pale, early-spring selves. I tell of calm mornings sitting on the porch shirtless with coffee while the hummingbirds hover. I tell of the movements of ravens, and the thrill of spotting an elusive red-faced warbler. I tell of how I've come to know and live Wordsworth's "calm existence that is mine / When I am worthy of myself." I choose in the final days to sleep on the cot in the tower overnight, so as not to miss the coming of the dawn. No one calls me on the radio anymore. No hikers appear. My time is my own and so are the moods of the mountain.

My walks now tend to focus on the near-at-hand. I trace the ridge to the west where it drops like stair steps off the peak. I hunt for the precise place on the east slope where the ground moisture becomes sufficient for the aspen to thrive. I circumnavigate the edge of the open meadow on top, which has a shape like a boomerang, the tower in its center near the apex. I visit favorite trees both living and dead, the biggest of the Douglas firs older than the founding of the republic, the fallen ponderosa stripped of bark and bleached almost white in the sun. The spring, reduced to a trickle in mid-June, runs in a steady stream again. Spider threads glint like delicate trip wires in the light of sunrise.

One evening I'm cooking dinner over the stove's blue flame when I look up and see, through the west-facing windows, two bull elk with their muzzles to the ground in the meadow. They are massive, majestic, the muscles in their hindquarters rippling as they shift their weight. One of them lifts his regal head and seems to look at me, his antlers stark against the gray sky; he shakes his jowls and returns to his grazing. I slip out the door and sneak around the corner of the cabin. When they hear me coming they look up, crouch slightly, then bolt, their hooves thundering down the mountainside. My blood races. Their musk hangs heavy in the air.

I TAKE ALICE ON ONE final hike to the north. We drop off the peak and follow a series of ridges and knolls through aspen and fir, locust and Gambel oak, tracing the Ghost Divide where the McKnight Fire made its easternmost runs. Lavender fleabane are blooming everywhere, Mexican silene too. On a windswept ridge I pause and sit, looking out over the Rio Grande Valley to the east, the deep Black Range canyons north and west—the latter clearly country for horse people or the hardy afoot, country to be humbled in. Among some stones on a ledge of the ridge I've tucked discoveries from my evening walks over the years: elk antlers, turkey feathers, snake skins, dried mushrooms, pieces of charred pine shaped like a woodpecker's head, a mule deer's pelvic bone, a bees' nest I found on the ground beneath a tree shattered by lightning. I don't remember when I started thinking of this place as a shrine to those I've loved and lost, but that is what it has become, a spot where I gather detritus from the living world, reminders of the transitory paths we trace on earth, memento mori. I do not visit often. I do not linger long. I add a mule deer antler, shed a few hot tears on the rock, then retrace my steps to the top of the peak.

Sunset brings colors to make a man tremble, colors without

names—names would only defile such colors. I sit in the tower mute as a stone. The light in its going, in its disappearing act beyond the Mogollons, does preposterous things to the clouds in the sky. An almost imperceptible breeze blows smoke my way off the few remaining fires, harmless little smokes burning at most an acre a day. For a moment to get beyond language, beyond words—no worries, no yearnings, nothing but the colors of the sky received on the retina and channeled to the brain by the optic nerve. The sweet smell of burning pine duff permeates the air. I sit in the tower mute as a stone.

The very last of the light rouses me from my stool, pulls me to the window like a miller moth to the moon, so close my breath fogs the panes. I turn and there again are the names of lookouts past, memorialized in pencil in all four corners of the tower. Names hinting at stories, names begging questions, things I'll never know: How did you entertain yourself on fogged-in mornings, Eddie Cosper? Were you sweet with each other in the moonlight, Kent and Deanie Carlton? Did you play cards in the off hours, Gary and Jerrie Ruebush? How many bears did you see, Carol van Kirk? Was that you who buried your empty beer cans in the meadow, Tuffy Nunn? What were the names of your fires, Gail Stockman? Did the wind drive you half mad, Bill Head? Were you ever in the tower for the sunrise, Jim and Deborah Swetnam? Did you hang a feeder for the hummingbirds, Fred Weir? What secrets of the mountain were known only to each of you?

The questions—unanswered and unanswerable. I stare at the endless dark north and west, the big wild, more than a thousand square miles unlit by a man-made light, and I let the questions go and think instead of a line from the poet Richard Hugo: *If I could find the place I could find the poem.* I have found the place. This is my poem.

ACKNOWLEDGMENTS

THANKS TO: MY EDITOR, Matt Weiland, who teased a book out of a slender diary of one fire season and whose editorial guidance was invaluable; the good folks at Ecco, who supported this book right from the start; my agent, Jim Rutman, whose patience and persistence proved legendary; M. J. Vuinovich, the original friend in a high place; Toby Cash Richards, for taking a chance on a greenhorn in the beginning; Dave Foreman, for sharing his voluminous files and vast knowledge on the history of the Gila National Forest; Thomas W. Swetnam, professor of dendrochronology and director of the Laboratory of Tree-Ring Research at the University of Arizona, for deepening my understanding of the Gila's fire history; Dennis Fahl, Jean Stelzer, Jim Apodaca, Willie Kelly, and Anthony James, for teaching me about fire on the Black Range; Jim Swetnam, for sharing stories of his years on Apache Peak; Stephen Fox, for correcting my grammar and helping me understand Aldo Leopold; Chris Adams, for sharing his knowledge of the Warm Springs Apache; Stephen Crook, librarian at the Henry W. and Albert A. Berg Collection of English and American Literature at the New York Public Library, for facilitating my access to the Jack Kerouac Papers; Les Dufour, Shane Shannon, Kameron Sam, Ricardo de la Torre, and Jack Doyle, for packing in my supplies all these years; Sara Irving, Razik Majean, Mark Hedge, John Kavchar, Jean Stelzer, and Rob Park, for teaching me how to be a lookout by example on the radio; Sue

239

Abel, Ellen Roper, and Brenda Hubbard, for helping me get paid without my ever showing up at the office; Alexandra Todd and Karen Latuchie, for reading and commenting on portions of the manuscript; Larry McDaniel, for enlivening my explorations of the Gila; and above all Martha, for friendship, adventure, forbearance, and love.

SOURCES

Abbey, Edward. *Black Sun* (New York: Dutton, 1971). Abbey's "fire lookout novel," an autobiographical story of a cranky wilderness hermit and the woman he loves and loses.

————. *The Journey Home* (New York: Dutton, 1977). In "Fire Lookout: Numa Ridge," a diary of the season Abbey spent watching for smoke in Glacier National Park, he notes that "the technical aspects of a lookout's job can be mastered by any literate anthropoid with an IQ of not less than seventy in about two hours."

————. *Abbey's Road* (New York: E.P. Dutton, 1979). The short essay "Fire Lookout" looks back on Abbey's four years in a lookout on the North Rim of the Grand Canyon.

Abolt, Rena Ann P. "Fire Histories of Upper Elevation Forests in the Gila Wilderness, New Mexico, Via Fire Scar and Stand Age Structure Analyses" (Master's thesis, University of Arizona, 1997). Abolt's study lays out the effects of fire suppression in the high-elevation forests of the Gila Wilderness and makes the case for allowing some high-intensity fires to burn once more in spruce-fir and mixed-conifer forest types.

Ball, Eve. *In the Days of Victorio: Recollections of a Warm Springs Apache* (Tucson: University of Arizona Press, 1970). Ball was a dogged collector of Apache oral histories, and this "as told to" narrative by James Kaywaykla is the fullest and most gripping account of what it was like to be an Apache child during the Victorio War.

Callicott, J. Baird, and Michael P. Nelson, eds. *The Great New Wilderness Debate* (Athens: University of Georgia Press, 1998). In this raucous anthology, environmental historians and philosophers beat defenders of the wilderness idea about the head with truncheons of deconstructionist academic-speak, while wilderness defenders reiterate the case for a basic

241

level of respect toward the nonhuman world. It was followed a decade later by a companion volume, *The Wilderness Debate Rages On: Continuing the Great New Wilderness Debate* (Athens: University of Georgia Press, 2008).

Chamberlain, Kathleen P. *Victorio: Apache Warrior and Chief* (Norman: University of Oklahoma Press, 2007). I drew on this sympathetic (and sometimes speculative) biography of Victorio for my thumbnail recap of the Victorio War.

Colegate, Isabel. *A Pelican in the Wilderness: Hermits, Solitaries, and Recluses* (London: HarperCollins, 2002). This is surely one of the most charming books ever written on solitude; Colegate writes in a lucid and discursive style about hermits ancient and modern, religious and secular.

deBuys, William. "Los Alamos Fire Offers a Lesson in Humility." *High Country News* 188 (July 3, 2000). An early essay on the disaster that was the Cerro Grande Prescribed Fire.

Egan, Timothy. *The Big Burn: Teddy Roosevelt and the Fire That Saved America* (New York: Houghton Mifflin Harcourt, 2009). Egan's book recounts with narrative panache the events leading up to, and including, the Big Blowup of 1910. I found here the quote from the *Idaho Press* about clear-cutting the northern Rockies as a defensive measure against wildfire.

Erdoes, Richard, and Alfonso Ortiz. *American Indian Myths and Legends* (New York: Pantheon, 1984). A valuable source for my discussion of Native American fire myths.

Flader, Susan. *Thinking Like a Mountain: Aldo Leopold and the Evolution of an Ecological Attitude Toward Deer, Wolves, and Forests* (Columbia: University of Missouri Press, 1974). As it says right there in the title.

Foreman, Dave. *Confessions of an Eco-Warrior* (New York: Harmony, 1991). The cofounder of Earth First! recounts his career in conservation and the major ideas that drove it—a must-read for eco-freaks, wilderness lovers, devotees of deep ecology, and redneck patriots of the fecund world. Foreman once told me that his greatest achievement as the Southwest field representative for the Wilderness Society in the 1970s was fighting to keep the boundary between the Gila and Aldo Leopold Wilderness areas at one mile; the Forest Service had argued for a ten-mile buffer between them, five miles on either side of the North Star Road.

Fox, Stephen. *John Muir and His Legacy: The American Conservation Movement* (New York: Little, Brown, 1981). One of the first books to make use of Muir's personal papers, it beautifully recounts the early history of American conservation and Muir's central role in it.

Gott, Kendall D. *In Search of an Elusive Enemy: The Victorio Campaign, 1879–1880* (Fort Leavenworth, Kansas: Combat Institute Studies Press, 2004). This monograph argues that the Victorio campaign is analogous to the so-called "war on terror," but since Victorio was never actually defeated by American troops—he was killed by Mexican militias—the comparison is perhaps unintentionally apt as a study in U.S. military failure.

Hugo, Richard. *The Real West Marginal Way* (New York: Norton, 1986). Hugo's essay "Some Kind of Perfection" contains the line of his I quote, "If I could find the place I could find the poem," and the fuller context is something I've considered many times as an impermanent caretaker of a permanent (by human time scales) landform: "Sometimes it seemed the place was more important than the event since the event happened and was done while the place remained."

Hurst, Randle M. *The Smokejumpers* (Caldwell, ID: Caxton Printers, 1966). An entertaining memoir of jumping fires on the Gila in the mid-1950s.

Kerouac, Jack. *The Dharma Bums* (New York: Viking, 1958). Kerouac and Gary Snyder (known here as "Japhy Ryder") wander the West Coast as early pilgrims in a rucksack revolution. Good times.

———. *Lonesome Traveler* (New York: McGraw-Hill, 1960). As it says right there in the title.

———. *Desolation Angels* (New York: Coward-McCann, 1965). This novel represents the truest account of his lookout days outside the pages of his journal, from which he borrowed heavily.

———. *Book of Blues* (New York: Penguin, 1995). Contains the dozen "Desolation Blues" poems.

Kittredge, William. *Owning It All* (St. Paul, MN: Graywolf, 1987). I took the quote about his dreaming of marrying a nurse from his terrific and timeless essay "Redneck Secrets."

Leopold, Aldo. *A Sand County Almanac* (New York: Oxford University Press, 1949). Still the one great American book for students of the natural world and a human land ethic.

———. *Aldo Leopold's Wilderness*, eds. David E. Brown and Neil B. Carmony (Harrisburg, PA: Stackpole, 1990). Brown and Carmony gather a fascinating collection of Leopold's early writings on the Southwest, which chart his evolution from gung-ho Pinchovian to first-rate natural historian. It appeared later in paperback, from the University of New Mexico Press, under the title *Aldo Leopold's Southwest*.

————. *The River of the Mother of God* (Madison: University of Wisconsin Press, 1991). See in particular Leopold's fascinating and groundbreaking essay, "Some Fundamentals of Conservation in the Southwest," in which he first discusses "conservation as a moral issue" at length.

Maclean, Norman. *A River Runs Through It* (Chicago: University of Chicago Press, 1976). Although the title novella sparkles as an incomparable gem of American literature, the other long autobiographical story in the book, "USFS 1919: The Ranger, the Cook, and a Hole in the Sky," is also first-rate and contains some lovely meditations on being a fire lookout in the early Forest Service.

————. *Young Men and Fire* (Chicago: University of Chicago Press, 1992). This meticulous reconstruction of the Mann Gulch blowup remains the only literary masterpiece ever written on the subject of American wildfire.

Manning, Richard. *Rewilding the West: Restoration in a Prairie Landscape* (Berkeley: University of California Press, 2009). I took his quote about the West's commanding views from this intriguing study of an effort to rewild a Montana grassland ecosystem.

Meine, Curt. *Aldo Leopold: His Life and Work* (Madison: University of Wisconsin Press, 1988). An exhaustive biography, Meine's book was an invaluable source of information on Leopold's Forest Service career and the evolution of his thinking about ecology, conservation, and land health. I relied on it heavily for all things Leopold.

Muir, John. *Nature Writings* (New York: Library of America, 1997). Collects the best of his work between two hard covers.

Nash, Roderick Frazier. *Wilderness and the American Mind* (New Haven, CT: Yale University Press, 1967). Revelatory when it appeared four decades ago, it remains the starting point in any syllabus of books about wilderness and American culture.

Oelschlaeger, Max. *The Idea of Wilderness* (New Haven, CT: Yale University Press, 1991). Among much else, Oelschlaeger's book explores the evolution in Aldo Leopold's thinking, from a strictly "imperial ecology" to a more "arcadian ecology," as he formulated his land ethic.

Pinchot, Gifford. *Breaking New Ground* (New York: Harcourt Brace Jovanovich, 1947). The memoir of the man who put his permanent stamp on the Forest Service, it contains the wonderful story about Muir and the tarantula.

Pyne, Stephen J. *Fire in America: A Cultural History of Wildland and Rural Fire* (Princeton, NJ: Princeton University Press, 1982). Pyne is the granddaddy of fire historians, and this was his first entry in the field. His section on fire in the Southwest was particularly useful, and from the books of his listed below I gleaned most of what I understand about the history and cultural context of wildland fire in America.

———. *Fire: A Brief History* (Seattle: University of Washington Press, 2001).

———. *Year of the Fires: The Story of the Great Fires of 1910* (New York: Viking, 2001).

———. *Smokechasing* (Tucson: University of Arizona Press, 2003).

———. *Tending Fire: Coping with America's Wildfires* (Washington, DC: Island, 2004).

Russell, Sharman Apt. *Kill the Cowboy: A Battle of Mythology in the New West* (Reading, MA: Addison-Wesley, 1993). A scrupulously fair-minded account of public-lands ranching in the modern West from all points of view, Russell's book offers the ranchers' dubious defense of their livelihood so I don't have to.

Scheese, Don. *Mountains of Memory: A Fire Lookout's Life in the River of No Return Wilderness* (Iowa City: University of Iowa Press, 2001). This is the finest addition to the literature of lookouts in decades, recounting Scheese's years as a fire watcher in the mountains of Idaho.

Snyder, Gary. *The Practice of the Wild* (Washington, DC: Shoemaker & Hoard, 1990). The passage of Snyder's about walking, which I copied into my commonplace book, comes from his essay "The Etiquette of Freedom." This and all of Snyder's works are beautiful meditations on wildness, ecology, humility, and the search for meaningful play and meaningful work.

———. *Mountains and Rivers Without End* (New York: Counterpoint, 1996). Snyder's poem "Things to Do Around a Lookout" can be found in its entirety here.

———. *The Gary Snyder Reader* (Washington, DC: Counterpoint, 1999). I found the the lines I quoted from Snyder's "Lookout Journal" in this grand collection of four decades of Snyder's writings.

————. *Back on the Fire* (Berkeley, CA: Counterpoint, 2007). See in particular the essays "Thinking Toward the Thousand-Year Forest Plan" and "Lifetimes With Fire."

Suiter, John. *Poets on the Peaks* (Washington, DC: Counterpoint, 2002). The photography alone makes this book worth the cover price, but it also tells the fascinating story of Kerouac, Snyder, and Philip Whalen working as lookouts and discovering Buddhism in the 1950s.

Swetnam, Thomas William. "Fire History of the Gila Wilderness, New Mexico" (Master's thesis, University of Arizona, 1983). Researched and written during the early years of prescribed natural fire, Swetnam's thesis used dendrochronology to show that, pre-1900, surface fires occurred as often as twice a decade in the ponderosa pine forests of the Gila Wilderness.

Thrapp, Dan L. *Victorio and the Mimbres Apaches* (Norman: University of Oklahoma Press, 1974). A fine account of the Victorio War, it focuses largely on the military maneuvering.

Truett, Joe C. *Grass: In Search of Human Habitat* (Berkeley: University of California Press, 2010). Truett, writing from the perspective of a man who makes his home just beyond the border of the Gila National Forest, makes the intersection of grass and human culture more fascinating than would seem possible.

Williams, Gerald W. *The USDA Forest Service: The First Century* (Washington, DC: USDA Forest Service, 2000). A useful overview of major events and periods in Forest Service history.

Worster, Donald. *A Passion for Nature: The Life of John Muir* (New York: Oxford University Press, 2008). A full-fledged biography of the original proselytizer for American wilderness preservation.

Wuerthner, George, ed. *The Wildfire Reader: A Century of Failed Forest Policy* (Washington, DC: Island, 2006). Among other treasures, this anthology contains a marvelous essay by Tom Ribe, "Fire in the Southwest: A Historical Context," which spurred my own interest in Southwestern fire history.